From the Horse's Point of View

Beyond Natural Horsemanship: Horse Training's New Frontier

ANDREA KUTSCH

Translated by Helen McKinnon

Trafalgar Square
North Pomfret, Vermont

First published in the English language in 2021 by
Trafalgar Square Books
North Pomfret, Vermont 05053

Originally published in the German language as *Aus dem Blickwinkel des Pferdes* by Franckh-Kosmos Verlags-GmbH & Co. KG, Stuttgart

Copyright © 2019 Franckh-Kosmos, Stuttgart
English translation © 2021 Trafalgar Square Books

All rights reserved. No part of this book may be reproduced, by any means, without written permission of the publisher, except by a reviewer quoting brief excerpts for a review in a magazine, newspaper, or website.

Disclaimer of Liability
The author and publisher shall have neither liability nor responsibility to any person or entity with respect to any loss or damage caused or alleged to be caused directly or indirectly by the information contained in this book. While the book is as accurate as the author can make it, there may be errors, omissions, and inaccuracies.

Trafalgar Square Books encourages the use of approved safety helmets in all equestrian sports and activities.

ISBN: 978 1 64601 060 8

Library of Congress Control Number: 2021935386

Illustrations by Cornelia Koller/Kosmos except p. 14 and the EBEC Pyramid (courtesy Andrea Kutsch)

Design by Lauryl Eddlemon
Cover design by RM Didier
Index by Michelle Guiliano, DPM (www.linebylineindexing.com)
Typeface: Helvetica Neue

Printed in the United States of America

10 9 8 7 6 5 4 3 2 1

Contents

1 INTRODUCTION
Retrospective ... 1

Longing for Unity ... 2
Conflict ... 4
Different Experiences ... 5
The World of Polo ... 7
The Age of Horse Whispering ... 8
The Stages of Horse Training ... 10
Scientific Basis ... 11

2 EVOLUTION
Horse Training from My Perspective ... 13

Stage 1—Classical Equestrianism:
Dominance from a Human Perspective ... 13
Summary: Stage 1 ... 19
Stage 2—Clicker Training: Conditioning from
the Predator's Perspective ... 20
Summary: Stage 2 ... 22
Stage 3—Natural Horsemanship and Horse Whispering ... 22
Summary: Stage 3 ... 28
Stage 4—EBEC: Science from a Horse Perspective ... 29
Summary: Stage 4 ... 33

3 EBEC PYRAMID LEVEL 1
Basic Physiological and Psychological Needs ... 34

Maslow's Pyramid of Needs vs. the EBEC Pyramid ... 35
Ethograms—The Body Language of Horses ... 40
A New Opportunity ... 42

What Is Chronic Stress? 51
Ways to Measure Stress 52
Physical and Mental Health 54
The Horse's Perception 56
The Horse's Memory 59
The Impact of Age 61
Stress Management 62
Summary: EBEC Level 1 67

4 EBEC PYRAMID LEVEL 2
Interspecies Communication as a Foundation 68

Communication Between Animals 68
Communication Between People and Horses 69
Acoustic Communication in Horses 72
Principles of Training with EBEC 79
Nonverbal Communication with Horses 80
The Horse's Responses to Communication 83
The Horse's Expressive Behavior 84
Ethograms: The Vocabulary Book of EBEC Training 88
How Do I Read the Ethograms? 89
Gestures of Full Body Expression 91
Gestures with the Legs and Hooves 92
Impressive Gestures 96
Facial Expressions 98
Expressions of Alarm 106
Expressions of Aggression 108
Suppression of Expressive Behavior 108
Expressions of Sensory Well-Being 111
"Snapping"—"Licking and Chewing" 112
Gestures with the Tail 115
How the Training Scale Still Applies 118
Summary: EBEC Level 2 121

5 EBEC PYRAMID LEVEL 3
Learning Theories as the Foundation of Successful Training 122

Problematic Training Stimuli 122
Reward and Punishment: Primary and Secondary Reinforcers 124
Communication and Behavior 125
Types of Conditioning 128
Classical Conditioning 129
Operant Conditioning 131
Aversive and Pleasant Stimuli from the Horse's Perspective 136
Distinguishing Between Innate and Learned Behavior. . . . 139
Summary: EBEC Level 3 156

6 EBEC PYRAMID LEVEL 4
Clear Objectives and Focus 157

Summary: EBEC Level 4 159

7 EBEC PYRAMID LEVEL 5
Fulfilling Potential Through Self-Reflection 160

Self-Reflection 162
Summary: EBEC Level 5 165

Afterword. 166
About Andrea Kutsch Academy (AKA) 169
Sources 171
Acknowledgments 173
Index 174

Introduction

Retrospective

Horses captured my heart when I was just a little girl. My parents didn't have any, but three Icelandic horses lived near our house. They were kept in an open barn setup that was very modern for the time and only ridden by their owners on the weekend. Every chance I got, I climbed over the fence and spent hours grooming and stroking them. I often used to just sit in the field and watch them, magically drawn to their steady breathing and even, predictable calm.

I'm a sensitive person; I was a sensitive child. I found many aspects of the adult world too hectic, too loud, too incomprehensible. I became nervous and anxious.

I experienced a predictable rhythm to life around these horses that only changed when they perceived a threat. I knew that the bird that had just taken off was nothing to be scared of. But I understood from the horses' point of view, the bird taking off was a potential threat from which they needed to first run away before weighing the risk and then settling back down.

When I was older I also enjoyed riding. There was another place in the neighborhood where you could rent ponies for less than a dollar an hour, but for me riding them wasn't really much fun. The ponies did what they liked. The people who owned the ponies gave those of us who rode whips and told us we should smack the ponies, something I couldn't manage with my small child's hands. I also think my hands just didn't want to do it—I loved horses as much then as I do now.

Almost every dollar I could get hold of ended up in the hands of these people. I usually just let the ponies graze as soon as we had left the barn. I didn't enjoy pulling up their heads by the bridle to stop them from eating and whipping them to get them to go forward. It never felt good. So, I paid my money and gave the ponies and me a treat.

LONGING FOR UNITY

At first, I just looked for the magical, calm, and harmonious world of horses—being part of the herd, a quiet place free of discord and raised voices. Horses radiated inimitable calm, strength, and magnificence that appealed to and attracted me.

People often ask me why girls feel so drawn toward horses. From a scientific point of view, I don't have a solid theory. For me personally, it was just that I felt confident and at ease around them.

Horses don't make demands, they just want to *be*. They eat, groom each other, look out for each other, and ensure their survival as a herd. On the many occasions I was grounded as a child, I felt excluded from the family group—a scenario I have never observed in horses. I always experienced them as a strong group with absolute unity. If you didn't threaten them, they didn't feel any anxiety about their survival or show any signs of unease or nervousness. That has always inspired me.

Human coexistence and life are considerably more complex and involve a lot more drama than that of horses, and that was what I felt as a child.

But I was at home with horses; with them I could be myself. I talked to them for hours, and because I could only ride occasionally, I didn't ask anything of them either.

I soon noticed that the horses' behavior changed when I wanted to do something they didn't like. If I wanted to take one of the Icelandics away from the others—into the stable to brush him, for example—he would dance around nervously, even break loose or squash me (*me*, who only wanted to be nice to him!) against the wall to the point where I couldn't breathe and got scared. The horse didn't want to be alone with me. He wanted to get back to his friends. These were times when I was in danger and just didn't understand the horses' behavior. The adults in my life said, "Horses are dangerous. Be careful!"

When I, a little squirt all of eight years old, had succeeded in dragging the horse behind me in from the field to the grooming area where I wanted to tie him up, just like the grownups had shown me, I wasn't able to brush him in peace. I would sometimes get annoyed because, from my point of view, the horse was spoiling my happy time. I wanted to spend hours grooming him and picking out and oiling his hooves, while all he would do was whinny for his friends. From my human perspective it was understandable that I got annoyed. Nobody thought about the horse's perspective back then. Horses were supposed to politely "toe the line," otherwise they were considered aggressive, unrideable, or dangerous. Horses were rarely able to shake off such labels. Adults said that horses mustn't be allowed to "get away with anything" and that you needed to assert yourself and "show them who's boss."

But when I watched an adult yell at a horse, I saw that it didn't make the horse more obedient at all, just more and more tense and anxious. I felt that something about that wasn't right. I couldn't bring myself to imitate such behavior. I just assumed it was because I was still too small, too young. But would things be any better later on when I had size and physical strength? I wasn't able to come to terms with what I saw in adults and what I felt was right inside at that time.

CONFLICT

Somehow, this inner conflict continued through the next 40 years of my life with horses. Do you need to assert yourself, use your authority, and make the horse submit to your will?

I have been on a long journey with horses. To this day, I have always maintained that we need to practice seeing things from the horses' point of view. We need to learn to understand their natural behavior and to find and offer appropriate solutions that they can follow without fear. Because, *without* us, *without* our actions and ambitions, horses don't have any problems at all. This point fascinates me.

But when I was a child, I wondered if horses just didn't like us. Were they afraid of people? Why couldn't we find common ground?

When all three Icelandic horses were standing together in the barn, before the whole group went out for a ride, I could brush and braid them for hours without any problems. I could do all these wonderful things that I enjoyed so much.

Sometimes one of the horses pawed the ground. I didn't like it, but it also made me think. The adults shouted at the horse: "Hey, cut it out!" Sometimes this would bring a few minutes' peace, but when the horse started pawing again, the adults often ignored or didn't even notice the behavior. Meanwhile, I wondered why the horse had pawed the ground, stopped, and then started again. When I asked the grownups, I was told to "stop asking so many questions!" So I learned to keep quiet, thinking it was because I was too young or too stupid so they didn't want to explain it to me.

There were just so many examples of horse behavior that made me think. When a horse bucked, kicked out, or was nervous, I thought, "Oh, the poor horse!" At the same time, I admired the adults who were able to assert themselves. I thought that one day I'd be able to do it, too: A smack, a shout, and all would be well. I was somehow convinced it was necessary and that adults knew what they were doing.

After all, I, too, experienced consequences. If I did something that was wrong from my parents' perspective, I was punished. If I stuck to the rules then everything was peaceful. If I refused to comply, I got into trouble. I was raised the way that horses were raised—the same principle, but sometimes maybe even harder. It seemed logical, consistent, and compelling. But sometimes it went wrong. A horse would break loose, struggle in fear, or fall over. The principle of punishment didn't always work.

I always found this type of exercise of power over another to be unpleasant. But I was told it was "necessary"—end of story! When I was young, if someone gave me a whip, I squeezed my eyes tightly shut and held my breath, because I felt so bad punishing a horse. I didn't want to make him frightened and upset. It felt wrong.

But I still did as I was told.

DIFFERENT EXPERIENCES

After my time with the Icelandic horses and the rented ponies, I was given a pony of my own to ride. I was totally "over-horsed" but made the best of it. Dominik, a German Riding Pony, was kept at a "competition stable," so I was pre-programmed to ride competitively. I hoped that I would finally get answers to my questions from the experienced riders, competitors, professionals, and instructors. My thinking was, "Well, one of them must know, because everything we do with horses has been thought up and passed on by somebody."

I followed the traditional and classical path of riding and training taught in Germany. It was okay as long as everything went well and the horses did as we said. But if something went wrong and the horses didn't do as we said, we punished them. I conformed, made an effort to succeed, and got more and more into dressage and show jumping.

But mostly, I loved preparing horses for competitions, looking after them, warming them up before, and cooling them down afterward. My own competitive ambitions paled in comparison, and I never got beyond competing at the lower levels. There was some kind of "block" in me. I always felt compassion if the horses weren't happy.

Yes, horses could be dangerous, so they had to be subordinate to people in a way, but in my heart of hearts, I disapproved of the approaches and many of the training techniques that went with this. I couldn't accept these techniques as "normal." After all, there were many other areas of life where shouting and hitting were now considered unacceptable. We never hit our dogs, and children aren't allowed to hit each other. I began to think that we, meaning all adults, just didn't know enough about horses. Many problems resulted from a lack of functioning communication between people and horses.

We didn't have answers to the horses' questions. I stored this knowledge deep in my soul. The time I spent more or less alone with the Icelandic horses as a child laid the foundation. I still remember how I would ask them, "Can't you tell me why you are afraid to go to the barn by yourself, without your friends?" I also talked a lot to my pony Dominik, asking him, "Why do you buck when I take you out on a trail ride? Why aren't you grateful that I've taken you out of your stall and given you a chance to get outside?" Neither Dominik nor the Icelandics answered, but that was only because I had yet to learn their language.

Dominik was eventually sold, and I continued my education as one of the best *Turniertrottel* or "TT," which is an affectionate German term for a competition groom.

THE WORLD OF POLO

I was playing polo at the time I turned to horse whispering. When training the polo ponies, the same questions from my childhood and the world of competitive riding kept coming up: "Why do horses sometimes not do what I ask them to do?" I always felt so bad when horses didn't work as they were supposed to.

Take a horse like Sundance. I bought the mare not because she was such a great polo pony, but because I had fallen in love with her. This was something that happened to me quickly and continually with practically every horse I had access to at the time. When a horse had problems I wanted to save him. Sundance was probably ten years older than I was told she was, but I didn't care. I was in love.

If Sundance heard the sound of the polo ball being hit she would immediately take off to the other side of the field as fast as she could. I would stand up in my stirrups and haul her around by the reins, to at least have a chance of taking a swipe at the ball at some point.

Polo is a game of strategy. The first player to strike the ball attempts to pass it to the players in front so they can score a goal. When I was riding Sundance, I always played in position No. 1, the goal scorer—an attacking, offensive position. There wasn't much to it other than riding fast and straight. When I tried to slow Sundance down, she bucked, so I only hit the ball very occasionally because most of the time I was already at the edge of the field while the others fought over the ball in the middle. It was pathetic. But Sundance was Sundance.

It was stupid of me to have bought her as a polo pony, but in retrospect she was the horse who gave me the key to unlock the answers I'd been searching for. I learned something when I was left unable to touch her for two weeks after Argentinian grooms had used their "educational measures" on her behind closed doors. And when I think back to a vet recommending that I sedate the mare before a game so she couldn't run as fast and I could get a chance to hit the ball, my blood still runs cold.

These people meant well, and at the time I clutched at every straw. But nobody could solve the problem. Not even a famous German rider who thought Sundance could be controlled with dressage training.

I was a reasonable show jumper, pleasure, and dressage rider, but it was obvious that I didn't know how to train a polo pony. "It's probably down to my lack of ability," I thought at the time. But I cried into my pillow more than once because I felt terrible that I couldn't understand what Sundance needed. Just like I did with the Icelandics, with poor Dominik, and with many other horses who I can't mention or this book would be thousands of pages long. They all have a special place in my heart.

I kept searching for the ability to understand horses.

THE AGE OF HORSE WHISPERING

Natural horsemanship and horse whispering gave me inspiration, or at least I thought they did at the time. Trainers who followed these methods renounced the use of a lot of gadgets. Instead, they used other, new aids that were unfamiliar to me at the time.

Above all, natural horsemen were the first to use body language in communication with the horse. The human—the "alpha"—copied behavior observed in wild horses. It was a fascinating discovery that was successfully applied and marketed by many trainers of this era.

The developers of natural horsemanship assumed that horses are able to decode gestures. The way in which horses responded to nonverbal communication with calm and understanding was a completely new field and fascinated me from the very first second I witnessed it. It began for me like it did for many, with the international blockbuster movie *The Horse Whisperer* starring Robert Redford. It introduced the idea of nonviolent, nonverbal communication with horses.

I set off on my journey, learned from the best at the time, and practiced this groundbreaking approach in the evolution of horse training down to the last and smallest detail. I mainly trained horses with issues and showed how I was able to mount untrained horses in 30 minutes and turn problem horses into calm and docile partners, in demonstrations watched by audiences of up to 20,000 people. I became known as the "German Horse Whisperer" who could solve problem behavior.

We horse whisperers had the solutions for horses who bucked, reared, refused to load, kicked the farrier, couldn't be caught or led, or were afraid of umbrellas. No matter how bad the problem, we had the solution. We could correct it by using the horse's nonverbal communication, imitating gestures, and applying consequences for unwanted behavior that we could take from the wild. We copied horses' systems of teaching each other, and they understood us beautifully. We saved the lives of thousands of horses and also turned working with horses into a kind of educational entertainment. It was an exciting time, and I was able to learn a lot.

At some point though, I wondered why, despite all this advancement of our understanding, horses were still often anxious and poorly behaved. The memories from my childhood remained, and I continued to ask, "Why don't you understand? Why are there problems?"

I didn't want to correct any more problem horses. I wanted to get it right from the start.

But how?

THE STAGES OF HORSE TRAINING

One day, a close friend Andreas von Veltheim gave me a copy of the good old *H. Dv. 12: Army Riding Regulation 12: German Cavalry Manual on the Training of Horse and Rider* that he had bought at a flea market. As I began to read, no one could have imagined the spark it would ignite. It marked the start of my research because it made me think about how horse training actually began.

I read numerous books, beginning my journey into a study of scientific works. The development of a science-based means of communicating with horses started with literature research and ultimately resulted in the diagram, "The Evolution of Horse Training" (see p. 14) and my concept of Evidence-Based Equine Communication (EBEC)—a training methodology designed from the horse's perspective.

I went through many stages in this evolution, and I found that each stage has one thing in common: it works from the perspective of the "predator," or of the human who devised it. The human isn't always wrong with his suppositions about how the horse could probably best understand human messages. However, it is often a system of trial and error. Sometimes it works, sometimes it doesn't, so it cannot be defined as a reliable method. The human has an idea, tries that idea and, when it works, that's wonderful. When it doesn't work, most people try something else that others have shown them. They often try lots of ways, many of which are unsuccessful. You might lose some horses, but when it works, it's great.

SCIENTIFIC BASIS

It became clear to me that we needed science-based results from the horse's perspective that shifted subjective human thoughts about what horses need into the background and facts into the foreground. There was an obstacle to overcome, however: I needed a team of scientists to help me put into practice my research concept for devising a scientifically sound method.

I began presenting my concept of EBEC at universities across Europe in 2005. At the end of a very laborious and strength-sapping series of presentations, I counted myself lucky to have a team of motivated scientists, universities, and investors contractually bound to me. From then on, the Andrea Kutsch Academy (AKA) had three partner universities: the Department of Veterinary Medicine of the Freie Universität, Berlin; the Vetsuisse Faculty of the University of Zurich; and the European University Viadrina in Frankfurt (Oder). On October 30, 2007, I applied for accreditation of the world's first Bachelor of Science in "Horse Communication, Riding, Training, and Teaching." AKA initially acted as a non-state university of applied sciences. I formed sections for business management and veterinary medicine (including reproductive medicine, complementary and alternative medicine, preventative healthcare, parasitology, surgery, orthopedics, animal nutrition, animal and environmental hygiene, applied anatomy, kinesiology, euthanasia, and pain research). Experts from ethology and psychology were also essential. My research team was enhanced by experts from classic and traditional teaching, farriers, riding instructors, and trainers from all equestrian disciplines, and consisted of over 40 professors and experts.

Paul Schockemöhle was another important partner. He provided us with his team and his stud of over 4,000 horses with almost 800 births per year at The Lewitz Stud in Neustadt-Glewe, Germany. More than 7,000 horses were made available to us over the course of our five years of research activity. On August 20, 2009, AKA was recognized as a state college, and in 2010, I was able to successfully complete my research concept for developing EBEC. AKA is now involved in global exchanges with numerous universities from different subject areas so its methods are continually enriched by the latest research results.

Evolution

Horse Training from My Perspective

My research into the evolution of horse training began with German and traditional teachings, and I came across plenty of other reading material from recent centuries that defined the training of riders and the training of horses.

STAGE 1 — CLASSICAL EQUESTRIANISM: DOMINANCE FROM A HUMAN PERSPECTIVE

The symbol in my diagram on page 14 for what I consider Stage 1 of horse training shows a horse lying down who is being dominated with various aids. The year "1884" is in reference to the book *The Gymnasium of the Horse* by Gustav Steinbrecht, which was published that same year. *The Gymnasium of the Horse* is still considered to be the timeless classic of equestrian literature, whose first edition was published by one of Gustav Steinbrecht's pupils a few months after the author's death. It also provided the basis for the *H. Dv. 12*.

Prior to that, the first military riding manual, entitled *Instruction zum Reit-Unterricht für die königlich preußische Kavallerie* [Riding Instruction

THE (R)EVOLUTION OF TRAINING HORSES

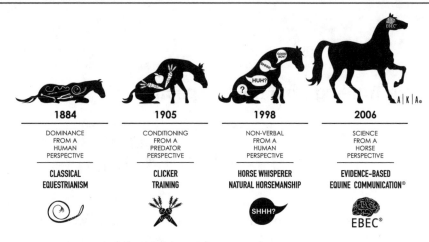

THE (R)EVOLUTION OF HORSE TRAINING
The four stages of the evolution of horse training.

for the Royal Prussian Cavalry], had been published in 1822. And in 1791, Ludwig Hünersdorf had brought out a book on training horses, meaningfully entitled *Die natürlichste und leichteste Art Pferde abzurichten* [The Easiest and Most Natural Way to Train Horses].

"Training Horses"

Hünersdorf analyzed working with horses from his human perspective, because at the time of his book there was still no scientifically investigated thesis on the nature or language of horses. He described behavior patterns related to the nature of horses that still apply today. His work paved the way for horse whisperers who, 200 years later, reflected on communication with horses and tried to further develop the classical and traditional methods.

Along with many other cavalry captains of this time, Hünersdorf was already thinking about the horse's natural impulses way back in 1791. It was something he wrote that prompted my own train of thought about the instinctive behavior of horses:

"However, the horse naturally has a strong desire for freedom that is only stifled by obedience, but its liveliness must still be preserved because, shortly before, the horse's gaits and his movements depended on his independence alone. This freedom stops the moment the rider takes up the reins, because from then on the horse is told what way he should go.

We should therefore limit the degree of subservience in the horse as much as possible, and change nothing about how he carries himself and his natural gaits, but content ourselves when the horse moves forwards when we ask him to." (Hünersdorf, 1992)

It was from this that, in 2005, I derived a reward system from the horse's instinctive behavior, following my extensive research of relevant literature. I then examined my reward system scientifically.

The use of coercive aids on horses who do not cooperate has been commonplace and long a part of riding instruction. As I've described, I learned early on that I should use my whip if my horses didn't want to jump. And if horses didn't respond to the reins—"their brakes didn't work," as my riding instructor at the time used to say—it was not uncommon for bloody foam from the harsh bit to drip from the corner of the horse's mouth. I forced myself not to cry for fear that I wouldn't be allowed to keep riding.

Hünersdorf describes that the horse must get used to being "ruled" by the rider and that the rider must learn to "defeat the horse with reason" and not with strength. He also conceded that when you make any changes, you must be satisfied with small amounts of progress. So, the art of small steps was already being taught in the eighteenth century. Even then, repetitive gestures such as head tossing, and therefore equine behavior patterns, were being observed, and the first attempts at describing them were being made. This philosophy is one that many horse whisperers and followers of natural horsemanship still adhere to today.

Punishing Horses

Even during this first stage of horse training, trainers admitted that people had to look for shortcomings in themselves first. Hünersdorf complains about riders who don't correct their horses properly. He wrote:

"They have hardly sat on the unwitting animal before they pull up his head unnaturally, and when the horse balks and won't go forward, they kick him in the ribs and rain down blows upon him. They think the harsher they are, the sooner the horse will be obedient. We regretfully see the innocent animal full of fear and confusion, and no good can come out of this kind of handling." (Hünersdorf, 1992)

On the other hand, he also describes when horses should be punished and with which consequences:

"In the case of hot horses, however, the rider should just resist with their hand so that the horse punishes himself with the bit at the most… also give a couple of really heavy blows from time to time…and let them know his displeasure about their inattentiveness in all kinds of ways." (Hünersdorf, 1992)

We find something similar in Stage 3—Natural Horsemanship and

Horse Whispering (see p. 22). With these training methods, when a horse's behavior is unacceptable, the human no longer punishes the horse for his behavior by hitting him, but by imitating how horses behave toward each other. For instance, a horse in a wild herd defuses situations by showing gestures of submission. So in natural horsemanship, the human might drive the horse forward in a round pen until the horse makes gestures of submission, at which point she will allow the horse to come in to her. Fear responses shown by the horse when he is being "driven" in this way have been proven to be significantly less pronounced than when the horse is punished by being hit or shouted at.

Hünersdorf also writes about the use of the voice in training:

"Because if the horse doesn't go forward enough in response to the tongue alone, and you really let him hear or even feel the whip, he will definitely be more attentive to the tongue next time." (Hünersdorf, 1992)

Impressively, an attempt was already being made here to successively intensify the consequences of undesired behavior. The trainer starts an exercise by making a noise with her tongue. If this doesn't result in the desired behavior, the trainer then uses the whip. The next intensification might involve the spurs.

Even a form of negative reinforcement—in this case, removal of the spurs—is described. A rider is supposed to rotate her leg so that the spurs touch the horse and cause him pain, but when the horse yields to the spurs, the rider should remove them from the horse's sides. This is basically a first attempt at using negative reinforcement (removal of a stimulus) as described in operant conditioning (a method of learning that creates associations between behavior and consequences by rewarding desirable responses and punishing undesirable ones—see p. 128). This was defined and examined for the first time in 1905, as shown in the diagram on page 14 with its reference to "clicker training," which traditionally employs a target, a "clicker," and treat rewards in order to achieve training goals.

However, it is clear in the calvary training manual *Instruction zum Reit-Unterricht* [Instructions for riding lessons] (Stritter, 1882) that training didn't always involve moderated amounts of pressure:

> *"If you have a horse that is so lazy, stubborn and insensitive that it shows almost no sensitivity to the rider's aids and even punishment, the horse should be tied between the pillars so that he can be whipped until he wakes up and his sensitivity is aroused."* (Stritter, 1882)

According to the *H. Dv. 12*, which superseded the *Instruction zum Reit-Unterricht* for the cavalry, the horse must not be punished in anger. However, punishment using the common aids, whip, and spurs, was an acceptable "means to an end":

> *"When it comes to punishment, the rider needs to be clear about whether the horse is really disobedient or whether he is raising the bar too high or has given unclear aids. The rider should never punish the horse in anger, because unfair use of the spurs that is only done in anger spoils the understanding between horse and rider and undermines the horse's trust in his rider. However, if punishment is necessary, the spurs must be used clearly and decisively. Half measures lack the intimidating effect on the horse."* (Sonntag, 2008)

We can see that riders were developing a sense of responsibility and were increasingly being asked to look for mistakes in themselves, as shown in the following passage:

"In most cases, disobedience is caused by the rider's incorrect effect and insufficient preparation of the horse. Horses also often use environmental factors as an excuse to escape the aids and therefore the challenge of flexing at the poll. The rider shouldn't just be intent on imposing his will but must seek to recognize and eliminate the true causes of disobedience." (Sonntag, 2008)

At this stage of the evolution of horse training, punishment is still deemed necessary, but we find repeated indications that it should be calm and dispassionate. It is also made clear that it is important that the horse is praised for even the slightest willingness to cooperate, because there is an agreement that reward and punishment are important means of communication between human and animal.

SUMMARY

Stage 1 in the Evolution of Horse Training

- Classical teaching based on the use of the horse for military purposes.
- Equine behavior is described and evaluated from a human perspective.
- Punishment and instruments of punishment are part of horse training.
- The horse's nature should be taken into consideration when training the rider.
- Punishment should be appropriate and meted out without emotion, so as not to lose the horse's trust.
- The horse's trust and nature are interpreted differently by every rider.

STAGE 2 — CLICKER TRAINING: CONDITIONING FROM THE PREDATOR'S PERSPECTIVE

The illustration that represents Stage 2 of the horse training evolution in the diagram on page 14 depicts a semi-recumbent horse who is being trained with motivation and food rewards. The year 1905 refers to the award of the Nobel Prize to Russian physiologist and behavioral researcher Ivan Petrovitch Pavlov.

We owe the second major phase in horse training to Nikolaas Tinbergen, Konrad Lorenz, Burrhus Frederic Skinner, Ivan Petrovitch Pavlov, and Edward Lee Thorndike and their experiments into instinct theory, classical conditioning, and operant conditioning. They differentiated between triggering instinctive, innate behavior that is not learned, and the conditioned, learned behavior of an animal who we can retrain or train in many different ways as desired.

Classical Conditioning

Ivan Petrovitch Pavlov (1849–1936) was a Russian doctor and physiologist. He was awarded the Nobel Prize in Physiology or Medicine in 1904 for his work on the digestive glands. His seminal work on behavioral research laid the foundation for behaviorist learning theories.

Pavlov assumed that behavior was based on reflexes. It was discovered that dogs in a laboratory salivated as soon as they saw a white coat. The white coat was initially a neutral stimulus. Whenever anybody wearing a white coat entered the room, the dogs were given food. This prompted the experiment that has become known as "Pavlov's Dog."

In the course of these investigations, Pavlov established that dogs' salivary secretion doesn't just begin when they start eating, but when they see food. He paired the stimulus "sight of food" with the previously neutral stimulus "sound of a bell."

A bell was rung every time the dogs were fed. After regular repetition of this process, the dogs secreted saliva and other gastric secretions when they heard the bell ringing. The formerly neutral stimulus "ringing of the bell" became a conditioned stimulus that triggered the behavior "salivary secretion."

Classical conditioning is based on the theories and the experiments that were done in the twentieth century. However, none of it was relevant to the world of horses at that time. Experiments were done on birds, dogs, and other animals, but behavioral research on horses didn't start until much later. It was initially about developing an instinct theory—documenting innate, unlearned behavior that occurs in every member of a species and that can be triggered by a stimulus.

Clicker Training

Clicker training is based on the assumption that an animal will show a specific behavior to get food. In this training method, the animal is given a food reward after he has demonstrated a desired behavior. This food reward is paired with a sound—the "click." The animal consequently understands the "click" as the reward.

However, there is doubt about the efficacy of clicker training in horses, because a scientific study was able to prove that unlike other animals, horses only produce saliva when they chew. Consequently, it can be assumed that presentation of food will not elicit any reaction in the horse. Food is always freely available to horses in their natural environment, so from their perspective there is no need to associate it with a specific behavior.

Therefore, clicker training *can* work with horses, but it doesn't always. It is in the realm of trial and error.

Instinct Theory

In my view, instinct theory and the behavioral biology (ethology) that later developed from it was a very important stage, and I am extremely grateful for the results of these experiments. Without the development of learning theories and their differentiation from instinct theory, it would certainly not have been possible for me to develop a scientifically-based system of equine communication. It would have been even less possible without the experience and valuable preliminary work of the classical and traditional trainers, the horsemanship people, and the many horse whisperers who endeavored to communicate with horses.

SUMMARY

Stage 2 in the Evolution of Horse Training

- First scientific studies into stimulus-response models and behavior protocols of other animals.
- Clicker training is a type of classical conditioning to a sound (click) that is associated with a food reward.
- General definitions for innate and learnable behavior are formulated.

STAGE 3 — NATURAL HORSEMANSHIP AND HORSE WHISPERING

The third stage in the evolution of horse training diagram on page 14 is represented by a horse who is almost on his feet. He is being trained with nonverbal communication. The year 1998 refers to the release of the movie *The Horse Whisperer.*

Ideas related to horse training and riding instruction were revised after World War II, and from then on there was more allusion to the nature of horses and how they communicate with one another. This was advanced during the era of horse whispering as more horse people conceived of new ways of training.

I'm convinced that natural horsemanship and horse whispering blew open the long-accepted ideas of traditional riding instruction and equestrian literature, especially since the spotlight was suddenly on so many "problem horses." I can't be the only person who wondered at the time, "Where do they all come from?" Tens of thousands of people packed venues to watch demonstrations that showed how to solve problems with horses. They were looking for help.

The more popular this training became, the more clearly I was able to see that we were working on the symptoms of problems, not the causes. There was plenty of room for interpretation of the reading material, plenty of scope for observation. Despite what seemed like vast changes in how riding and training were taught, punishment in equestrianism today still involves a harsh voice, whip, and spurs, and rewards come in the form of a quiet voice, patting, and feeding. Basically, not much has changed. But in my everyday work with horses and people, I have seen a lot of freedom in how natural horsemanship techniques are interpreted, leading to confusion among horse owners, as well as trainers and riding instructors.

When I began my scientific research in 2006, lots of people were searching for "the truth"—for solutions *from the horse's perspective*. But despite the wealth and variety of information, there was no groundbreaking *scientific* information about equine behavior to be found.

I suddenly realized that all of the natural horsemanship approaches were still mostly concerned with human interpretations and that there was a lack of actual scientific evidence about why horses do what they do and how they communicate.

"Violence-Free Communication"

Back then I was promoting violence-free communication with horses, but I noticed that this was an empty cliché that was interpreted in a different way by every person who heard it. It took a long time for me to realize that. For one person, a smack with a whip isn't violence because it "wasn't doing anybody any harm." For somebody else, a smack with a whip was an animal-welfare issue. A tug-of-war began over basic definitions.

Official literature from equestrian federations and organizations began to contain information that horse whisperers used. However, it was mainly only conceded that horses are flight animals, herd animals, and herbivores. By and large, competitive and professional riders paid no attention to the methods advocated by natural horsemanship.

As an example, consider how the FN's *The Principles of Riding* describes how you should "be quiet," "keep a lid on your anger," and "praise" the horse. But how exactly this is supposed to be done is largely left up to the reader to decide. Artificial aids such as whips and spurs are still permitted in both "traditional" riding and natural horsemanship. Whips are commonly used either as "an extension of the arm," or as throughout history, as a tool for punishment. Admittedly, there is a clear attempt to involve the horse more during this era, but tools that cause pain are still sometimes used.

If a rope causes the horse pain when he puts his head down to buck and he stops bucking, we can assume that his behavior changed because the rope triggered pain. If a chain pulls tight on the horse's nose when the horse is wound up or anxious, it might cause him to quiet because it hurts. Putting iron rings around the horse's fetlocks to prevent him from pawing the ground, might work because every time he does so, the metal hits his fetlock, causing discomfort.

Shows, Expos, and Public Arenas

All of the above are examples of techniques used in the third stage in the evolution of horse training. It is definitely the best-known time frame because in the eyes of those outside natural horsemanship, it seemed like "mystical things" happened. There was suddenly knowledge, or more precisely, *ideas* about how to use body language, an aspect of communication to which no or very little specific reference was made in the first and second stages of the evolution of horse training. Problem horses could now be transformed into "nice" horses with the use of almost invisible elements, namely body language. Horses who had for years refused to load into a trailer could be loaded up in minutes.

But it didn't work every time.

I can still clearly remember introducing a horse to his first halter, saddle, bit, and rider in just twenty minutes, in an arena, watched by an audience of 10,000 people. The crowd went wild. The horse was as sweet as the many hundreds of other horses I had trained in public. However, I got a call a few days later saying that there were problems. The owners hadn't been able to get on him since the demo.

I wondered why the information in the horse's brain hadn't reached his long-term memory. After all, my rider had legitimately ridden the horse in walk, trot, and canter—no smoke and mirrors. The horse was so good that day in front of the crowd that I was moved to tears myself. So how could this be?

I became the victim of the power of subconscious messages because it was easy for me to put the blame on somebody else—in this case, the owners. I'd learned from my traditional German instruction that "those who didn't grow up around horses" or "those recreational riders" don't have a clue! I can no longer remember exactly what information my subconscious was feeding me. Nobody, neither the teachers I'd had from my traditional equestrian roots nor those I'd learned from in the horse whispering scene had thought, "Hmm, interesting. Why hasn't the horse remembered our message? We need to look at what we can do better." Until then, I hadn't come across anybody who had answers to my questions. The simple explanation was always that it was somebody else's fault. It was always somebody else. Yep, the owners probably didn't have a clue.

But the thought, "Why didn't the horse remember?" wouldn't let me go.

The horse from the demonstration had been incredibly easygoing. I would have taken him out on the trail at home after the one session we'd had, full of trust and confidence. The owners seemed to be kind. They hadn't frightened the horse because he was quiet and free from anxiety. They hadn't confused him either. He was just fundamentally good and well raised. So why couldn't they get on him at home?

I was also struck by the horses I found wouldn't load after being perfectly happy to do it in a public demonstration in front of hundreds or even thousands of spectators. These same horses would trudge up and down the ramp, sometimes up to 10 times, without so much as batting an eyelid, without whips or raised voices—without any voices at all, just with body language and a loose rope. It was a mystery to me why after the demo was over there would be a horse who just about had to be shoved into the trailer. Of course, this didn't always happen—there were countless horses who proved to be permanently "healed."

And then there was the extreme. Similar to Stage 1 of the evolution, when the horse accepted the leg, good. If he didn't accept the leg, a whip and spurs were used until he gave the desired response.

As I said before: sometimes it worked, but sometimes it didn't.

A Method?

I feel quite unsettled when I look back. I will never forget one horse who we worked with at the State Stallion Depot of the Stud in Neustadt (Dosse), Germany. He was supposedly "trained," but he acted like an untouched stallion who clearly had issues. Someone had been on him in the past, but from what we knew, had not managed to stay there.

It didn't matter. We wanted to train him to saddle. Pride also played a role. It was a nightmare for me. The horse was wild. He immediately became sweaty and upset. Our rider couldn't stay on him. So, we somehow managed to tie a dummy to his back. We fastened it so securely that it couldn't fall off. And the horse tore around, bucking circle after circle.

A giant of a horse. Big, athletic—a breeding stallion. The professional riders with me said that since the traditional, classical tools hadn't worked, the dummy was now the only option. No rider could stay on that horse, but he could only get the dummy off with great difficulty. In my view the horse was done. He was fighting for survival.

I couldn't seem to manage to work out how negative reinforcement was being used in accordance with application of learning theories. The horse might carry the dummy eventually, but still no rider would be able to stay on him, of that I was certain. The technique wasn't just dangerous, it was also ineffective. My thinking was "it isn't much different from what the earlier trainers did, the sweat and the struggle, restraining the horse between pillars with chains and beating him."

After discovering that the training I'd done with the horse in the demonstration wasn't successful after the owners had him at home, thoughts flickered through my brain: *If a method doesn't work on every horse, it isn't a method. And if nobody knows why it doesn't work, then we don't know enough.* My mind was on a rollercoaster, and my thirst for more knowledge strengthened.

Looking back, the sweet horse from the demo was an important horse on my path to development, including *personal* development. I wanted to develop further with regard to horses, but also when communicating with people. To do so, I had to set out and replace what I suspected to be true with actual knowledge. Stage 3 was a time of trial and error, and just as valuable to the evolution of horse training as Stages 1 and 2. It was the first attempt at copying the body language that horses use among themselves to enable humans to become equal partners. Punitive actions based on those natural to horse herds were used to make horses submissive.

For me, the development of learning theories became the most exciting part of my own education. Once I had studied equine learning theories in their entirety and done detailed research into the literature, my journey could really begin. I wanted to combine all three stages of the horse training evolution together, and maybe even find a little inner peace. Peace within me, as a grown woman. Peace with my pony, Dominik, who I know now was badly harmed by the training methods of Stage 1. Peace with the Icelandics who pushed me to my limits due to my ignorance. And peace with Stage 3 when I became Germany's most famous horse whisperer who was suddenly depended on to help all the misunderstood horses.

But there was one thing that wouldn't let me go: Just where was the common denominator?

SUMMARY

Stage 3 in the Evolution of Horse Training

- Natural horsemanship and horse whispering are based on observations of the behavior of wild horses and an attempt to copy this behavior.
- People try to imitate horses and become "horse-like" themselves.
- Trainers copy the "alpha" mare in the herd and try to win the horse's trust.

STAGE 4 — EBEC: SCIENCE FROM A HORSE PERSPECTIVE

In 2006, I cut the ground from under my own feet. I had become known and established in Germany as "the horse whisperer," but I knew that I was missing something. I could feel it with every horse. There was still room for improvement, and I was far from knowing everything I needed to.

A Mare Named Seductress

My beloved mare Seductress helped me to understand horses on a new level. She was a Thoroughbred, bred for racing, and for me she was the sweetest, most trusting horse who, like so many others, found a special place in my heart. It was love at first sight. I wasn't her owner to begin with, but her trainer.

I prepared her for her first saddle, rider, riding itself, the racetrack, the starting gate—for everything that she might encounter in her life as a racehorse. I wanted to make sure that she wouldn't cause any problems if she ended up at a racing stable where she would be expected to perform well within a short time. Straightforward horses are always popular at racing stables. This is very important, especially in a professional training stable. When there is something wrong, the horses are often labeled "nut jobs," and then there is trouble when the jockey sees on the list that his is supposed to be riding the "sack of sh*t." Horses with issues don't get off to a good start. I wanted Seductress to be sweet and kind so riders would try to help her, and so she would experience less pressure, violence, and misunderstanding.

The significance of the name or the "stamp that a horse has on his forehead" is a psychological phenomenon. According to scientists, our brain subconsciously takes care of 90 percent of everything we do without us being aware of it. Our brain does routine work by itself. Well, riders form opinions about horses, categorize them according to their experiences—their own or others'—and judge them within milliseconds. They (subconsciously) give them labels, whether they mean to or not. It's how the human brain is structured.

Experiments have been done where subjects are poured the same wine. Some of the subjects were told that the wine was a particularly expensive and good vintage. These subjects' brains categorized the wine as higher quality than those of the others who were not given any information.

Our opinion is easily influenced, and we make the same kind of judgments about horses.

The information that Seductress would give her rider would determine how her future would be. I did everything I could think of with her to make sure that she had the best possible chance at a good and well-loved life at the racing stables. I loved her with all my heart, but like I said, she didn't belong to me. All I could do was train her to the best of my ability; the rest lay in the hands of the owner and the racing stables.

My plan worked. Seductress was well liked and cooperated with everyone. But her upright forehand conformation couldn't cope with the demands of racing. Her legs couldn't withstand the strain of the high speeds that racing requires.

After just a few starts and at just two-and-a-half years old, she was lame—"broken down," as the professionals say. I was there when the training stables phoned and said, "She's no good!" They had taken all possible veterinary steps, but Seductress didn't have the conformation required for the racetrack.

I can still feel how my heart started beating faster at the thought that she would be coming home to me. I thought that I would get to love her and discover all of the other great things she could do.

I can still remember how speechless, how shocked I was, when the owner said: "She's no good. We're sending her away." I wondered what he meant by "away." It didn't take long for my eyes to fill with tears at the shocking realization that "away" meant a "one-way trip." A trip to the slaughterhouse, no return. I can still remember crying and quietly asking the owner: "Is it always about money? She is your horse after all!"

I knew that racing was all about money. In the end, money is always the yardstick for everything that we do with horses. Nevertheless, it still came as such a shock to me because I knew that this owner had believed in violence-free training and had been an accomplished follower of natural horsemanship. How could that be?

I obviously understood that racing is a business, but I wanted to find another way for Seductress. I quietly asked in a trembling voice: "Would you give her to me?"

And so, Seductress became my horse. The friendship with the owner fell apart for me, even though we stayed in touch for years. I began to really sit up and take notice, and that is why I owe it to Seductress that I began to develop the EBEC Pyramid (see p. 36). I hadn't yet begun to study the science of horse behavior, but I was becoming increasingly involved in scientific subjects. I now understood how traditional riding instruction had come about. I had heard the stories of punishment and reward and the use of whips and spurs. The understanding I gained meant that I no longer heavily criticized their usage. They are part of the definition of riding instruction; they were there at the foundation of equestrianism. I began to observe the riders and the tools, more with the wish of recognizing success when it happened, without assessing what was being used and when. I thought that way I might be able to discern a system.

Seductress was the best, kindest, and most wonderful horse that I had ever owned, but she wasn't physically able to cope with training. So I asked new questions: If a horse isn't 100 percent fit to compete, can he be trained anymore? If not, what does that actually mean? Can we always tell whether a horse is "healthy enough"?

And I was thinking not just about the horse's body, but also about his mind.

Too Much Stress

In 2006, people said that proof had been found that horses excreted stress hormones when they were sent onto a circle. This spoke to me. This means that sending a horse onto a circle as a negative consequence for undesired behavior puts a horse under stress.

If that is the case, then it would mean that, according to the EBEC Pyramid (see p. 36), my horse wasn't even able to comply with my aims, in exactly the same way that Seductress wasn't able to carry a trainer or jockey to victory because of her conformation, no matter how hard she tried. Why? Because to be able to focus on actual learning, horses need to have a calm, focused mind *without* stress, and *everything that we ask of them* is something that they have to learn. They have to learn to wear a halter, because this is not something they are familiar with in the wild or born with in domesticity. They would never wear a saddle, climb into a trailer, or stand in a small stall of their own accord.

If we look at the entire system of equestrian training against the background of the first stage of the evolution of horse training, there is compulsion, sharp words, whips, spurs, drilling, and submission. Actions that cause stress in the horse have an extremely detrimental effect on his ability to learn. The learning theories from Stage 2 were developed from the predator's perspective and are consequently not directly transferable to the horse, who is a prey animal. The Pavlovian experiments were done on dogs, and equine clicker training was in some cases copied from dog training, so these methods came not from the perspective of horses, but that of dogs.

Learning isn't possible under stress and the application of learning theories will fail or rather won't reach the horse's brain under stress. Therefore, if they are to be able to learn, horses need to not experience any stress. The task before me now was to investigate how horses respond to rewards and punishments *from their perspective*. I believed we need to reach horses with clear communication. We needed a science-based training method that, for the first time, considered the horse's perspective. The five levels of the EBEC Pyramid (see p. 36) had to be brought to life.

> ## SUMMARY
> ### Stage 4 in the Evolution of Horse Training
>
> - EBEC is based on scientific investigations into instinctive behavior and the horse's brain.
> - A completely new reward and punishment system was developed that focuses on the horses' perspective and is derived from their instincts.
> - EBEC prevents problem behavior, noticeably keeps horses free from stress, and prepares them as well as possible for everything that is expected of them in equestrian sports and riding.

EBEC Pyramid Level 1

Basic Physiological and Psychological Needs

In order to be able to learn, a horse needs to be healthy, calm, and receptive. What does that actually mean? Based on Maslow's hierarchy of needs (see p. 35), often depicted as a pyramid, it is essential to guarantee the horse's basic needs to enable him to concentrate and absorb information as well as possible.

People use horses in a whole variety of different ways. For example, I just wanted to be allowed to keep loving Seductress, while her racing trainer wanted to see her succeed on the track. I wanted to groom and spend time with the Icelandics, while their owner wanted to ride them on trails and enjoy the great outdoors. My colleagues at the training barn and fellow grooms wanted to have a job so they could earn money. The riders from my youth wanted competitive success to boost their egos and to gain recognition as successful or professional riders.

All people have horses in their lives for different reasons, and they also deal with individual horses in different ways. Horses have to cope with this and make the best of it. They try to assert their needs for protection, safety, to be part of a herd, and their survival instinct. But the different objectives when it comes to humans and horses can lead to conflict. The good news is that while horses only have a limited ability to think strategically and are not able to cognitively classify our human needs, we humans are able to learn to see things from the horses' perspective. This allows us to make our lives with them calmer, more harmonious, and, therefore, more successful.

Key Points
- Humans place many demands on horses that don't correspond to their natural behaviors.
- We need to be aware that in order to be able to cope with these demands, horses are in a constant state of learning.
- For horses to be able to learn optimally, they need to be calm and free from stress.
- Horse owners must make sure that their horse's basic needs are met.
- We can only succeed in doing this well by seeing things from a horse's point of view.

MASLOW'S PYRAMID OF NEEDS VS. THE EBEC PYRAMID

Maslow's hierarchy of needs pyramid is a motivational model that deals with the type and effect of human motives. In his 1943 paper, Abraham Maslow put together the following five levels:

1 Basic needs that are necessary for preserving human life, such as breathing, water, food, sleep, reproduction, shelter, sexual behavior, and maternal love.

2 Safety and security needs—needs such as physical and mental security, basic financial income, work, accommodation, family, health.

3 Social needs like friendship and connection that interestingly only arise in people when the first two categories have largely been satisfied. Only then will somebody experience a strong urge for social relations. This is also called *affiliation motivation*.

4 Maslow counted trust, esteem, self-affirmation, success, freedom, and independence as individual needs.

5 Self-realization: According to Maslow, if all of the needs prior to this level have been satisfied, a new restlessness and dissatisfaction will arise in people. They will want to display their talents, potential, and creativity; develop further in terms of their personality and skills; and shape their life and give it meaning.

Maslow's hierarchy of needs cannot be transferred to the needs of horses. Social needs, for example, manifest themselves very differently in horses than they do in people. It would therefore be a fallacy to assume

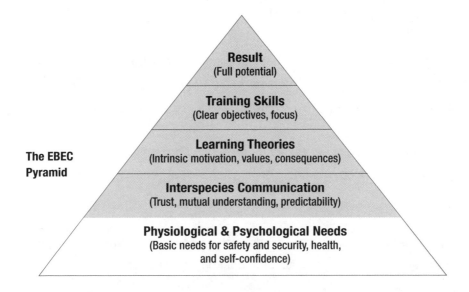

The EBEC Pyramid

Result
(Full potential)

Training Skills
(Clear objectives, focus)

Learning Theories
(Intrinsic motivation, values, consequences)

Interspecies Communication
(Trust, mutual understanding, predictability)

Physiological & Psychological Needs
(Basic needs for safety and security, health, and self-confidence)

that horses have the same needs as people. This book would miss the point by thinking that it was. The point is for people to learn to adopt the horse's perspective—the horse's point of view—based on scientific facts. Nevertheless, it is still crucially important not to leave our human needs out of the discussion. We can only change our traditional methods, deal with other living beings in an empathetic way, and adopt their perspective if we understand our own actions.

It's about developing more willingness to accept the fact that a horse's behavior will show us if his basic needs are not being met. This can be done through behavior that people don't want, such as pawing the ground when being groomed. A lasting change in that behavior—that is, learning—can only be achieved through a training process. That is what Evidence-Based Equine Communication (EBEC) strives to make possible.

Different Basic Needs

In Level 1 of the EBEC Pyramid we ensure that the horse is calm and free from stress. His basic needs for protection and security are met: Shelter, water, grass or hay, and as a herd animal, social contact with familiar herd members. He feels "free," even if the field is fenced, and he has a high feeling of self-assurance as he wants for nothing. He can lie down to rest and stretch out flat so that REM sleep relaxes the structures of his brain. He feels secure in his environment and doesn't show any physical signs of anxiety or unease. All in all, he is calm and content.

Understanding Learning Is Not Possible Through Punishment

It is not possible for horses to learn through "punishment" alone. Let's take removal of water, as basic need, as an example. First, the removal of water could be harmful to the horse's health. If a horse won't go into the trailer, you could take away all water sources, except for a bucket of water in a trailer in the field. Yes, the horse might go into the trailer in search of water, and the trailer might lose its negative connotations over time because it is the only place with water. But this doesn't work every time. If the horse is then forced into the trailer at the end of the week when he is also thirsty, he could end up being afraid of trailers for life and will be anxious whenever he has to go in one. Going forward, whenever he arrives at a competition, this horse might not perform as well as other horses who arrived in a state of calm and with their basic physiological needs satisfied.

EBEC (see diagram, p. 38) does not involve any "trial and error" techniques that produce stress or trigger anxiety. This will enable the horse to perform to the best of his ability and to be ready for all of the needs he is supposed to satisfy *in us*. The human will then be happy, and the horse will be happy, too.

Survival Instincts

Every living being has constant and enduring needs, the strongest of which will assert itself over the others. Horses are no different in this respect. Their survival instinct will assert itself, especially if one of their basic needs isn't being satisfied and their survival is therefore threatened. The horse might then behave aggressively—kicking, biting, bucking—as he "defends" himself from us. In these cases, people like to say: "You need to show him who's boss!" But that is an old-fashioned view that has been disproven by scientific study. No, we don't have to "have it our way" every time and dominate the horse. We *do* need to ensure that all the horse's basic needs are being met. And as mentioned, above all, the horse must be free from stress.

Many times, a training plan will be created from our perspective, not recognizing that it is completely against the horse's nature. For example, from our viewpoint, there is absolutely nothing wrong with riding out alone on the trail, without the horse's "herd." However, according to the hierarchy of needs of the horse, this is absolutely *not* okay. With patience and understanding, the horse can learn to do this task that is a "human idea," and when he does, he will find that his survival isn't actually threatened.

Priorities

Over time, Maslow expanded and revised his model, and he established that some needs have priority over others. For example, we need air and water more urgently than we need a new car. A horse needs feed appropriate to his level of work, somewhere safe to sleep, and social contact with other herd members more urgently than he needs a new trailer or new leg wraps.

According to the book *Motivation and Action* edited Jutta and Heinz Heckhausen, a need will activate and influence action only for as long as it is unsatisfied. Action is less "pushed" from the inside than "pulled" by the consequences of satisfaction. Therefore, the motivational power of a need decreases as satisfaction of this need increases. For example, you don't try to drink more when you stop being thirsty. When a horse is under stress—that is, stress hormones are being released—he, as a flight animal, won't eat. When that stress is relieved, he will graze again.

There are very significant differences between the natural behaviors of horses and those of people, especially between those that people believe they can make strategically. Science describes the human cognitive brain as a "public relations campaign for subconscious action"; 90 percent of all our actual actions are based on the experiences stored in our memory. It is common for us to assert ourselves and sometimes to "dominate" or "oppress" others—especially animals, and in this case, horses—because this is what we have learned to do or because we think it is right. However, if we really want to, we can work on ourselves to improve our natural behavior, and this book is supposed to encourage us to rethink such actions when it comes to horses.

ETHOGRAM — THE BODY LANGUAGE OF HORSES

In order to simplify interpreting equine behavior, we have drawn up *ethograms*—an inventory of behaviors or actions exhibited—that make the body language of horses readable and generally comprehensible (see chapter 4, p. 88 for more about these). This leaves as little room as possible for human guesswork.

For some people, a horse is considered quiet and calm when he can comfortably sleep in a barn among other horses, is fat, and moves so slowly he can hardly put one foot in front of the other. For others, a horse is quiet and calm when he doesn't smash up the stable walls when his neighbor leaves. Some people might consider the horse to be quiet and calm as long as he isn't sweating, rearing, pacing, or cribbing in the stable, and regularly say things like: "He has his little moments!" "That's normal!" "It's just the way he is!" or "Hey, cut it out, you psycho!" However, others might judge similar situations in a completely different way.

Much equestrian literature tries to create awareness that people should behave toward horses according to their nature. However, until EBEC came into being, there was no standard definition of how exactly this should look.

For example, in 1998, the FN (German National Equestrian Federation) explained why problems with horses or problems between people and horses might occur:

"It should also be taken into consideration that we are increasingly dealing with horse people who come from an urban environment and who are therefore not as familiar with the natural needs of horses and/ or other animals than was the case in earlier generations. It is therefore

no longer enough to just teach 'technical skills.' Special value must be attached to knowledge about the horse."

And by 2002, the "experienced horseman" was still described by the FN as somebody who "knows"—but what that person knows exactly is not described and defined. As a result, there is still far too much scope for interpretation:

"The experienced horseman can correctly interpret the different characteristics of equine expression and behave accordingly, especially with stallions and wild horses. Horses also behave differently around different people. This can vary from trusting devotion to a person to maximum aggression. The different kinds of behavior must be recognized not only when dealing with the horse on the ground but also under saddle. Being familiar with them is an important prerequisite for avoiding accidents."

The empty clichés, and therefore, rather meaningless messages that the reader or listener receives in these teachings are problematic, including this one:

"The better the rider can feel and think their way into their horse, and the more skillful they are, the more rarely they will get into problematic situations with their horse."

Who sets the standard for "feeling"? What exactly is "feeling your way in?" How is this defined by experienced horsemen, horse whisperers, followers of natural horsemanship, and even plain old horse lovers?

I became aware of these issues when I started to think scientifically about horse training and worked to draw up hypotheses. Science forced me to be specific. I learned to observe without judging and to refrain from empty clichés. It was incredibly difficult at first, because I had to reconcile my new way of working with my life up to then, as a classically trained, traditionally taught competitor, polo player, pleasure rider, and horse whisperer.

But it was a eureka moment. Changing perspective—seeing things from the horse's point of view—is considerably more difficult and complex than it sounds.

From then on, blanket interpretations became taboo. My thinking had to become fact-based. It was healing, in a way, and caused me to reflect on myself, which is so important for getting to a turning point in the search for solutions.

A NEW OPPORTUNITY

From the horse's point of view, it would definitely not be pleasant to refuse a water jump out of fear of the unknown and then be dragged through it by a tractor over and over, at night, until you stop putting up a fight. Hünersdorf had already clarified in 1791 that horses do not usually refuse out of *"ill will, but pure surprise and displeasure at what is happening to him."*

People will do unthinkable things out of desperation when it comes to satisfying their own needs. Acknowledging this enables me to be empathic and to understand somebody else's actions. But now, we all have the opportunity to change our behavior and do our best for horses by sticking to the facts and acting accordingly.

It was unfortunately the case, as eyewitnesses report, that there was a horse who was indeed dragged behind a tractor through the water jump at night and then actually jumped the jump in the Hamburg Derby the next day. However, if I were to teach this and write it down as a training technique, then we would be in the realm of trial and error, because such a method could also cause a horse to be afraid not just of tractors, but also, and even more so, of water jumps and arenas.

Creating measurable parameters eliminates these mistakes. The answer lies in the ability to adopt the horse's point of view and to prevent the release of stress hormones at the right time, in order to increase focus and the ability to learn. Sustainable learning will not take place if focus and learning ability decrease because the horse has gone into survival mode and is only functioning in the context of total overstimulation or overload at one moment in time. It is good to practice precisely defining the signs of stress to make observation possible on a factual level. This helps us along the way, not just in dealing with horses, but in our daily coexistence with each other. This book, and the ethograms I share in chapter 4, should help us to find common denominators and adopt the horse's perspective, a perspective that is measurable independently of our personal experience with horses.

When is a horse anxious? When he gives signals that are different from his normal, relaxed state. Nowadays we are familiar with these signals. We can correctly classify them, and thanks to modern scientific methods, we can say a little more about what the horse will do next or which hormones are active in the metabolism and have a beneficial or blocking effect on training. Ethograms give us a generally applicable and scientifically founded tool that helps us to apply learning theories in a way that enables the horse to optimally store information about the behavior desired by people.

It didn't take many scientific studies to confirm that an anxious, unstable, and from his own perspective, not fully receptive horse has a reduced capacity for learning that does not noticeably improve even after numerous repetitions or punishments. According to this substantiated finding, it is initially a question of defining measurable parameters for showing how horses look when they are stressed. What measurable or recognizable guidelines can we make available to help people act in a horse-appropriate way in these moments?

Horses Are Not Inherently "Bad"

The problem with most horsemanship guides is that the reader can interpret the written word in many ways. I've demonstrated this with some of my examples related to the evolution of horse training in chapter 1 (p. 14). In addition, every person who is involved with horses interprets horse behavior in a different way, according to the person's own experiences. It is, therefore, very important that we agree on a "common denominator" and accept something that has been scientifically examined and proven: Horses fundamentally do not mean us any harm. The EBEC Pyramid (see p. 36) clarifies this common denominator for us and the horse.

The equine nervous system processes signals that affect the horse incredibly quickly, especially on an emotional level. The human nervous system tends to act more on a strategic, planning level. Horses need to act quickly to ensure their survival. One wrong decision can have fatal consequences. When confronted with an external stimulus, horses have to decide whether to leave the situation—take flight—or to stay and wait to see what happens, focus, and attack if necessary. They have the option of driving away the stimulus or the person influencing their behavior.

Emotionally relevant external stimuli that have an effect on the horse have been scientifically proven to also have a major direct influence on the release of hormones controlled by the brain. These hormones then have a lasting impact on how the horse will behave in similar situations in the future.

For example, if being loaded into a trailer makes a horse nervous, there is an emotional reaction that causes stress hormones to be released. The process of hormone regulation generates a learning experience in the horse's emotional brain, which can then be responsible for the horse's future defensive behavioral responses. Examples of these behavioral responses could be a spook when just being led past a trailer, even if the human has no intention of loading the horse. The moment the horse sees the trailer, his brain makes a decision that triggers the behavior. This behavior often then provokes a lack of understanding in people, perhaps because our focus is somewhere completely different, so we often miss the important initial physical expressions of the horse's anxiety. When we keep our human perspective, we might feel the horse is being unreasonable if he suddenly leaps or dances around us, snorting nervously as we try to pass the trailer as quickly as possible. We may blame the horse and punish him with a jerk on the rope, a shout, or a smack. This sets in motion a mutual spiral of anxiety because the horse's emotional brain misses nothing.

We humans often find it difficult to shift our internal perspective to the horse and to see the world from his point of view, especially in moments that we initially thought were safe. When a horse suddenly exhibits a defensive behavior for reasons we can't understand, we must learn not to accuse him of "being bad," but to instead understand that his is a nervous, hormonal reaction rather than a cognitive behavioral response.

Horses can't see ghosts, but they can see or perceive things that we miss. This sympathetic basic understanding helps us to respond empathically to the horse's behavior. Instead of getting angry, we can decide to begin a training process that has a positive effect on the horse and that covers different experiences, without causing further stress that will make our life together increasingly difficult in the long term.

Horses Don't Scheme and Strategize

Horses have absolutely no interest in acting against us or planning things that could harm us. The structure of their brain doesn't even allow them to process such complex interrelationships. From our perspective, we often think that horses deliberately don't do things: "He doesn't want to go for a ride because he's lazy," or "He snaps at me when I tighten the girth because he doesn't want to be ridden," or, "He knows that he can get away with it," or, "He's ignoring the aids because he can't be bothered to do a flying change."

These are human trains of thought. From the horse's point of view, these scenarios are far less complex: "I canter and she kicks me in the ribs." The horse probably isn't ignoring the aids but instead doesn't know that he's supposed to do a flying change. He can't think it through strategically. Only when the horse is able to understand the triggering stimulus—in this case, the backward movement of the rider's leg that is supposed to precede the flying change—and that the flying change itself is associated it with this stimulus, will he then be able to put it into practice.

The basic prerequisite for training is always that the first level of the EBEC Pyramid is fulfilled. We must always keep these parameters in mind and review them constantly. The rest of the levels won't work if Level 1 is not satisfied. Review, and if necessary, correct the parameters. Causing the horse pain or fear via hitting, shouting, or kicking in this learning process will not escape the notice of the horse's emotional memory, and the next time you ask for a flying change at "X," you might find that your horse just rushes due to anxiety and in an attempt to escape the situation that he has already experienced and that his limbic brain has stored as unpleasant.

If several unsuccessful attempts are made even though Level 1 needs are being met, the problem might also be found in Level 2 or 3 of the EBEC Pyramid (see pp. 69 and 122). However, you can be sure that when all the levels have been satisfied from the horse's point of view, he will always do what is asked of him.

Be Understanding

Responding with understanding to the horse's behavioral responses, such as fight or flight, is not just of key importance to his general ability to learn but to ours, too. Everything else we have tried has proven to cost time, money, and in the worst cases, our own personal safety or that of others, leaving us to deal with the problem for longer. Punishing or confusing horses can therefore also have negative effects on us.

If we respond with aggression or anger in the heat of the moment, either out of a lack of understanding, awareness, perception, or self-control, and accuse the horse of "scheming" or "being bad," we can often create long-term problems. This has been proven in examinations of the hypothalamus in equine sexual and defense rituals.

The Hypothalamus and Amygdala

The *hypothalamus* is defined as the most important area of the brain for maintaining inner calm and for physical adaptation in stressful situations. Even a very small unpleasant external stimulus creates alarm signals. Imagine the hypothalamus as a control center that decodes everything that comes in from the outside.

If the hypothalamus has stored a stimulus as "good" and "safe," then the parts of the horse's body that are visible to us remain neutral. However, if it decides to categorize an event or stimulus that the horse has perceived as dangerous, a release of hormones into the bloodstream will immediately trigger a defensive or flight response. This is because of the neural connections between the hypothalamus and the other areas of the brain. The hypothalamus controls the release of hormones, and it is crucially important for us in the application of EBEC. If we can manage to positively influence the horse's hypothalamus, such that it responds to us positively because of actual positive experiences, this has an enormous impact on the entire relationship between horse and trainer, and on the horse's capacity for learning. The horse then "trusts" us more and more, especially in critical situations.

In reality, he doesn't decide to trust us with his strategic brain. Instead, it happens when we have continually fed his emotional brain with the correct information. We can create trust by giving the horse positive experiences. This means we can shape our future together by dealing with the horse consciously, in every situation that he questions, and by handling his behavioral responses to external stimuli in a way that causes the horse to lose his fear of the unknown.

Regardless of whether the circumstances are the way we would like them to be or not, they happen. The horse has a feeling about an object or event, and we have to deal with it—but we have the choice to be understanding. Horses work according to the principle of, "What did I do and how did it work out for me?" The more positive experiences we give the horse's memory, the greater his "trust" in our actions, and then we, horse and human, experience pleasure in each other's company, as well as successful training with very low levels of stress hormones.

The *amygdala*, the emotional brain, acts when a situation occurs that the horse finds stressful. Depending on how frightening or emotionally charged the horse assesses the situation to be, it is processed either directly via the sympathetic nervous system or via the hypothalamus.

In the case of the direct and rapid route, which is responsible for horses' incredibly fast reaction times of less than one second, the "Danger Alert" (*unfamiliar object in the ditch on the side of the road*) is sent to the adrenal gland, causing it to release adrenaline immediately. This increases the horse's pulse rate, even though he is not currently doing any physical activity. Investigations with heart-rate monitors have shown that increased pulse when adrenaline is released can be immediately distinguished from physical activity. This shows that the peak in release of the hormone is caused by stress.

However, the heart rate isn't the only thing that increases. The horse's blood pressure also immediately increases, make the muscles tense and hard. This is the rigid tension in the horse that we can instantly feel beneath the saddle with our seat, legs, and hands when we ride past a scary object. Because we can directly experience, measure, recognize, and feel this, in this situation, we can decide what information we send back to the horse in response. We also decide when or whether to respond with a rewarding or punitive stimulus. Do we give feedback and influence the horse's reactions as soon as his neck muscles become tense, his head comes up, and his gaze becomes focused (see ethogram on p. 101) in response to an unfamiliar object, or do we only begin to influence the horse when he has already started to spin or is running back to his stall, looking for herd members to protect him.

In this case, the horse's emotional brain has chosen the flight response—the behavior has already been triggered and now belongs to the past. We should not punish the horse's behavioral response once it has happened, because as far as the horse is concerned, it is already over and done with. It is more effective to notice and influence the *possible* behavioral response *before* it happens, when the horse's brain is still in the process of making a decision.

If we sharpen our senses, we can react *before* a situation becomes dangerous for us or for the horse. When we want to consciously be in control rather than be victims of trial and error, and when we also want to consciously feed the horse's brain important, future-oriented information, then we need to respond—not frantically, but as quickly as possible—to what we have consciously recognized and observed *from the horse's perspective* and without our own interpretation.

Ethograms help us spot these moments and correctly time our responses to them. The aim is to recognize early in the horse's behavior when his state of being changes from "neutral" to "concerned." The human should ideally respond when the horse is just moving into this state of concern, rather than one of "panic" or "uncontrollable fear."

Doing the Right Thing in Difficult Situations

It is important to be cognitively alert and self-reflective when handling horses. Situations with emotionally-charged horses can be frightening, especially when we don't know how we should respond in the situation. This has nothing to do with us not knowing what to do because we are new to horses or because our grandfather wasn't a horse person or because we haven't "earned our spurs" or because we are just pleasure riders, or just show jumpers, or just dressage riders, or just this kind or that kind of trainers. No, the knowledge is new and it will change life with horses forever.

How it looks when a horse responds to stimuli and which circuits are active in the brain that we can successfully influence has now been defined. The old cavalry captains didn't have this knowledge, and neither did the newer horse whisperers. So how much experience someone has with horses doesn't matter. Now *everybody* who deals with horses can come together to reference a standardized vocabulary of equine gestures. These gestures allow us to determine basic rules for successful handling and training, which lead to improvement in performance, all based on horses' natural behaviors. EBEC helps to immediately diffuse every situation in the company of equines because it gives you knowledge that enables you to adopt the horse's perspective.

We *know* that "the thing in the ditch" is harmless, or at least we know that with the cognitive part of our brain. However, if we are unable to switch on the cognitive part of our brain because we are already being controlled by the emotional part of our brain, then the thing in the ditch looks different. This is what it is like for the horse.

The horse's eyes, ears, or other senses have told him that the unfamiliar object in the ditch is dangerous. Adrenaline causes the horse's muscles to become tense, which we can feel and usually also see. Blood sugar supplies the horse's muscles with the energy that they need to be ready should the horse need to choose fight or flight.

Once it has put everything in the horse on alert—which it does in a fraction of a second—the emotional brain, the amygdala, now also informs the hypothalamus. The hypothalamus receives information about a possible need for action to preserve life and releases additional hormones that the blood circulation carries to the adrenal gland that then releases the stress hormone cortisol into the blood. At this moment, the horse is incapable of thinking about the person who is perhaps currently sitting in the saddle. The rider is out of contention, because considering the rider would require the ability to think strategically and to empathize and sympathize with her.

WHAT IS CHRONIC STRESS?

The cortisol that is released when the horse is put on alert is an important hormone for survival. However, constant presence of cortisol in the blood circulation is harmful to the horse.

Physical responses to mental or physical strain are described as stress. This can occur in dangerous situations where the body increases its readiness to perform. The respiratory and pulse rate increase, as does blood flow to the muscles. The horse is preparing to react as quickly as possible. When these stressful situations are a rare occurrence they are not harmful to the horse's health. Chronic stress, however, is deadly and should be avoided at all costs.

Chronic stress occurs when the horse is unable to find a solution for overcoming his anxiety in stressful situations that occur repeatedly and over a long period. Examples would be a broodmare who experiences claustrophobia in the stocks whenever she is inseminated, a horse who panics every time he is loaded, or a horse who is hysterical on the longe because he is afraid of the whip.

Repeated frightening situations cause chronic stress in the horse. Positive experiences, on the other hand, teach the horse how to deal with these trying situations and free up space in his brain for new information. This is another reason why we should avoid stress as much as possible when training—it both helps horses learn quickly and keeps them healthy.

In a nutshell, this means it is best to respond competently to the physical signals that the horse gives with his movement and expression when he first indicates there is a potential danger lurking in the bushes. We mustn't get angry when we notice that something in the horse has changed (his muscles have tensed, his head has raised), but instead analyze the environment from the horse's perspective and try to perceive where the unfamiliar stimulus is for the horse at that moment. We can then use this information at a later time to draw up a strategic training plan and to take meaningful steps to give the stimulus a positive association in the horse's brain, which will help him to remain calm and attentive when he perceives it again. When we don't do this, stress levels will only increase and influence how information is stored in the horse's brain in the long term.

WAYS TO MEASURE STRESS

We need to act as soon as the horse switches from "normal" to "excited" in relation to the first level of the EBEC Pyramid. Some of the signs are changes in:

- **Respiratory Rate:** We can make out how many breaths a horse is taking per minute by observing the movements of his flank as he breathes in and out. Although all values can vary from horse to horse, a basic general guide is that a calm, healthy adult horse takes around 8 to 14 breaths a minute. Foals can take 20 to 40 breaths per minute at rest.

- **Pulse (Heart Rate):** Level 1 of the EBEC Pyramid says that my horse must feel safe and be physically and mentally healthy and free from pain. The horse should have a normal respiratory rate when I bring him out of his stall, and also a normal pulse, which is around 32 to 36 beats per minute in an adult horse.

- **Body Temperature:** The horse's normal temperature is between 99.5 and 101.3 degrees Fahrenheit. With our horses in training at the Andrea Kutsch Academy (AKA), we often take and record their temperature, two to three times a day. As with the other baseline parameters, we can only influence this value when needed and in a timely fashion when we know that it is unusual.

Observe how your horse breathes normally at rest so that you can then compare during appropriate work or physical effort. It is worth noticing and remembering how it feels when everything is "normal" with your horse. When you know what is normal, you will notice more quickly when something changes.

Our biggest challenge in identifying baseline parameters and any changes is the horse's rapid reactions. This takes a little practice but can be fun once you know what you're looking for. When I'm training a horse and asking him to increase his pace, the respiratory rate changes according to the effort that the horse is putting in. It isn't about observing individual breaths or constantly riding with a heart-rate monitor (although this can be sensible with sport horses, so you know exactly when you are training aerobically and anaerobically); it's more a matter of knowing what is normal for the horse so that you can notice immediately when it changes.

Of course, all of the elements included in the individual levels of the EBEC Pyramid need to be considered together. If a horse and rider are schooling over a course of jumps, the rider will probably not be able to tell in enough time that the horse is going to refuse a fence from his breathing, but she may from the horse's movement and gestures that are defined in Level 2 of the EBEC Pyramid (see chapter 4, p. 70). For example, it can be helpful to pay attention to where the horse is looking when riding into the arena. Riding a circle at the right moment can prevent faults or a refusal. When I'm training and the horse's breath and heart rate show that he is calm, and I pick up an even rhythm in walk or trot and his muscle tone feels normal underneath me, then I'm prepared to notice any changes that might impact his ability to learn what I am teaching him sooner, with better results.

This is a good tip for riders of all kinds, whether they are competitive or pleasure riders. We assume that our horses are fit and healthy and that they look forward to working or spending time with us. But when we train or exercise horses, they depend on us to know whether they are capable of being ridden or driven, according to our needs. What we ask for sometimes requires a lot of basic fitness, so it is good to have measurable parameters to ensure activities are safe for the horse. This practice can also enable you to spot many of the early signs of metabolic disorders and illness, such as laminitis or colic, making it more likely the horse can recover without lasting effects.

PHYSICAL AND MENTAL HEALTH

It's always good to know what I can ask from a horse in training. Peak physical fitness gives me as a trainer much more confidence when it comes to maintaining or advancing the training plan. It will be more likely my horse will be able to physically and mentally reach my goals for him. The horse should be sleeping and eating well, and should not be lame or unbalanced. And we mustn't misinterpret physical signs of discomfort and judge the horse from our human perspective, thinking, for example: "You're just pretending because you don't want to work!"

Horses have been proven to be incapable of such thought processes. I know that this can sometimes be difficult for us to believe because it is human to plan our behavior in advance and to be able to feign pain. Horses are not able to do this. They do not have the prefrontal cortex dedicated to executive function that we do. Their first option for acting in frightening situations is determined by instinct. We, as people, need to learn to accept this and to train ourselves to deal with horses from their perspective if we want them to be contented, successful, healthy, and capable.

When a horse's pulse and respiratory rate increase the arteries dilate and the muscles become tense. The muscle tone changes and makes the horse tremble, appear unsure, and rock backward, as if prepared to take off. You can sometimes even hear the horse's teeth grinding—often described as a bad habit, this is more of an important indicator of the horse's inner state. (Note that using equipment and gadgets in an attempt to conceal a behavior like teeth grinding is a serious mistake with long-term consequences that will interfere with the basic harmony you wish to establish with your horse. An example would be a tight noseband. As trainers, use of gadgets moves us to the realm of trial and error. If it goes wrong, the choice you make can have dire consequences for the rest of the horse's life. The behavior can end up being displaced and other unwanted behavior can come into play, such as the tongue hanging out, loss of relaxation and suppleness, rhythm and gait issues, tension, high head carriage, and defensive or aggressive behavior.)

Physical unease can also be observed when the corners of the horse's mouth are tense and his face is fixed. His mouth will be dry and his cheeks will appear sucked in. This appearance signals a horse who is getting ready to make a decision his life might depend on.

It is in these little gray areas that we can have the opportunity to have an effective influence. We already know that the horse's decision for or against us or for or against the exercise that we are doing isn't made in the horse's cognitive brain. This is the most important thing to tell yourself when it comes to looking at the situation from the horse's point of view: "I accept the scientific evidence that the horse is not trying to trick me into not taking him out for a ride because he is lazy and would rather stay at home. No, he is behaving this way for a valid reason."

THE HORSE'S PERCEPTION

If we look at the horse's field of vision, it becomes clear that their all-round vision gives far more information than we are able to take in. It is simply a fact that horses see or don't see objects and movements that we see or don't see. As the person sitting on the horse's back, you can decide in every situation whether you react in a kind and solution-oriented way or aggressively, therefore increasing problems in the long term. Either way, the horse's sensory brain makes a decision about behavior the instant it spots that potentially dangerous object in the ditch. The amygdala receives information from the sensory organs—eyes, ears, skin, and nose—and it takes this information seriously.

The human knows that the woman with a dog on a leash is politely waiting on the path for them to ride past, but the horse might be unsure. If the human responds aggressively—hitting the horse with the whip, shouting, or jabbing him with the spurs, and taking the attitude of, "You can't let the horse get away with it, you have to show him who's boss!"—then the scenario will be registered as another source of stress that directly affects the horse. That's because the hippocampus, which is in a lively exchange with the amygdala about everything that happens, will remember the situation.

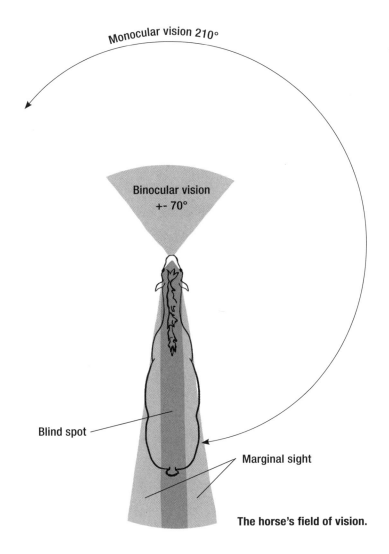

The horse's field of vision.

Like the amygdala, the hippocampus is part of the limbic system and can also be described as a central switching station. It plays a role in everything to do with filing memories as short-term and long-term. The hippocampus is also considered to be a part of the brain that consolidates memories that are then stored in various different places in the cerebral cortex. If the experience with the dog at the edge of the path was bad, the hippocampus will "remember" this stressful situation, and going forward will report a warning signal even more quickly than before. Any even remotely similar situations will instigate a similar behavioral response in the horse, even if in a completely different environment. The hippocampus will instantly and rapidly send warning signals to prevent the horse from getting into such a dangerous situation again. The horse is geared to spot danger early in order to keep stress levels as low as possible, prevent release of adrenaline or cortisol, and keep blood-sugar levels low and muscle tone relaxed. That is the horse's preferred state of being. So, when a similar situation occurs at a later date, the horse's memory will make sure that it escapes or raises the alarm even earlier. Let's say that it isn't the woman with the dog at the edge of the path; now it's something "different," but from the horse's point of view, a "similar-looking dark thing," this time in the barn aisle or in the corner of the arena. Let's assume that an upturned bucket in the aisle provokes a similar reaction; it could also be a cooler hanging over the fence in the arena. The process really gets underway when the human is once again too late to notice the horse's growing uncertainty related to these objects and decides to respond to it with punishment or aggression. The horse again experiences punitive consequences for his natural flight response. This produces an increasingly insecure horse with a greatly impaired ability to think.

Consider again the situation with the woman and dog at the edge of the path. If you respond from the horse's point of view and keep your distance instead of approaching, enabling the horse to recognize the woman and the dog as harmless with his binocular, or if applicable, monocular vision, the amygdala will send the hippocampus different information. The defensive response or anxiety behavior will not be as strong the second time a similar situation occurs—an upturned bucket in the aisle or a cooler over the fence—and positive repetitions will cause the defensive behavior to continue to decrease.

It's better to get it right from the start!

There are other benefits: This also applies to riding over a course of jumps or into a dressage arena or out on a trail. A horse who has been trained with EBEC and scientifically sound basics will be more trusting when he sees "other things at the edge of the path." The horse's behavior will become increasingly calm if subsequent experiences are repeatedly positive. Just remember: From the human's perspective, we often think things are small and inconsequential ("He shouldn't make such a fuss, it's only a bucket."), but from the horse's perspective, it's "something dangerous that must be overcome." This is why it is very effective in the long run if the horse can experience a potential threat as a positive "learning opportunity." He will then store the situation as an event that he has successfully overcome. Something that he thought was dangerous turned out to be less dangerous or not dangerous at all, and the horse was able to cope well with the situation. "Phew!"

The horse's long-term memory is then fed positive information that in turn boosts his self-confidence and self-worth.

THE HORSE'S MEMORY

From the horse's perspective, he is always encountering things that he rarely has a chance to understand and grasp. These are usually things that are harmless from our perspective, because we can understand and assess them.

It is therefore of utmost importance that we feed the equine system the right information. When we apply EBEC we welcome the horse's physical communication with us, because we can then make out his "internal map," and we know which information he has already stored as positive or negative. In order to be able to enter into dialogue with the horse, we need information from him about which triggers destabilize his system.

This is what makes the first level of the EBEC Pyramid so important. In order to be able to know and recognize when the horse's behavior changes, we have to recognize what is normal for him when there are no stress hormones in his system. Only then can we help the horse record and store a situation as harmless, even though it might at first have seemed dangerous to him. If the horse spots a "danger," then it is a danger from his point of view, and we can say, "No, it isn't," until we are blue in the face, but it will still be dangerous as far as the horse is concerned. Neither horse nor human can order their emotions to disappear, and we can assume that horses are a lot less able to do this than people, because they are not capable of putting themselves in a self-meditative state where they can "think over" the situation.

To be honest, even we can't order our fear to vanish if we think there is a bear rustling about in the bushes, because the adrenaline has already been released into our bloodstream. The fear has been triggered. Our emotional brain has initiated self-protection without any cognitive involvement on our part. We can't even order our fear to disappear if our companions reassure us that there aren't any bears in the area. Think about it: It would be much more helpful for us in such a situation if our friends responded with understanding, rather than shouting at us, "Stop making such a fuss over nothing!"

When I was a child, people often told me, "Stop making a fuss!" My system sounds the alarm as soon as I hear that, even today, because I wasn't ever "making a fuss." I was reacting because of how I felt in a situation, even if that seemed absurd from the adults' point of view. A lack of understanding in this kind of scenario can cause fear to escalate. Imagine being afraid of bears and then being smacked, shouted at, or left behind alone when you thought you heard one in the bushes. It could leave you in such a state of fear and panic that you might never want to go hiking again, even with other people.

This is not a cognitive decision. In the case of doubt, your subconscious will have excuses ready to explain why you can't go hiking next time. Of course, we humans have the advantage of being able to recognize this kind of subconscious pattern, and we can later overcome fears such as this with the help of a therapist. Horses can't go down this route, though. They can't say that they didn't realize that the water jump is actually just a piece of blue plastic with a little puddle in it. They are also not capable of reasoning that if they *did* fall in, they would at most make a splash and that their life is not at all in danger.

THE IMPACT OF AGE

It can be assumed that an excessive amount of stress in young horses will influence future stress responses. Information related to what takes place in frightening situations is stored in a way that ensures stress hormones are released more quickly and in greater quantities in the future. This is an irreversible effect that lasts a lifetime. Horses who have experienced very stressful situations during training as youngsters—an accident when being loaded, rough "breaking in," forced flexion of the neck to attain a certain headset, violent situations where they were "trapped" and unable to successfully escape—will be more susceptible to stress for the rest of their lives. They are more easily made tense, quicker to bolt, or more likely to try to break loose from a handler than those who have had positive experiences or whose first experience of a stressful or traumatic situation was in their adult life. This is extremely important information for us to recognize: It is in the horse's younger years that he learns all the basics that will accompany him for life.

I have been able to discern very clear differences in the thousands of sport horses I have trained and started, including during my years of study at Paul Schockemöhle's Lewitz Stud in Neustadt, Germany. I have done numerous blind experiments with horses who had received good, fear-free basic training with the application of EBEC. Long-term observation has shown them to be more successful in sport than horses who had been worked in traditional ways. So, while we would do well to train horses from their perspective in all situations in life, it is especially beneficial during early training. In the long term, the EBEC foundation is incredibly important for improved behavior and heightened composure in the horse, including in new and unfamiliar stressful situations that the horse might later encounter in competition.

The key is then, avoid stress in training and in situations where the horse is frightened. Help to understand rather than further frighten the horse or suppress the behavior with force or violence, or by ignoring it. How this works will become clearer as we look at the next levels of the EBEC Pyramid in coming chapters. Let's use our human, strategic abilities to adopt horses' emotional perspective and to help them to better cope in our world.

STRESS MANAGEMENT

Development of stress depends on the horse's highly individual perception of a certain situation. His individual perception depends on his genetic predisposition, the experiences he has already had and stored, and his sense of being able to influence the situation.

Individual Training of the Horse

Training horses means getting them used to unfamiliar things, which means triggering stress. However, this stress must be released in a controlled way to enable the horse to show and compete.

Every stimulus that we apply as a punishment, reward, or to reinforce a behavior when applying learning theories must be comprehensible for the horse, and the horse must be able to change it with his behavior, because horses, just like people, react differently in stressful situations. Something that is more stressful for one horse might be less stressful for another. I often find this to be the case when I am training many horses at once or bringing one horse after the other out of the stable or in from the field. When I am teaching several young horses to wear a halter, one might be really worried and keep pulling his head away because he can't accept the halter straps over his nose or across his poll. On the other hand, it might be possible to put the halter on a second horse for the first time without any problems, even though both horses have the same breeding and have been managed in the same way.

Being restrained goes against the nature of horses. We therefore have to work very hard to achieve the desired result: It should be possible to put the halter on the horse, leave it on, and take it off again without any defensive behavior or worry on the part of the horse. However, if the horse performs a defensive behavior, like head shaking or even rearing or kicking at the halter when he realizes that he can't get it off his head, the stress produced can be excessive. The problems I described previously can then arise and ultimately lead to us no longer being able to touch the horse in the stall with a halter in our hand and the horse turning his hindquarters toward the door and threatening to kick. Learning is a complex process and there are many things in its interrelationships that we still do not understand today. For that reason, seeing horses as individuals is crucially important in EBEC. Even if I am only training ten horses, it is rare for one horse to be just like another.

I recently worked with a horse who was still worried about coming out of the stall and going into the indoor arena with me by himself, even after four days. Meanwhile, the other horses from his age group had already been long-reined in the area with a saddle on without any problems. His fear of being on his own, of leaving his friends behind in the barn, was so great that he psychologically couldn't cope for more than 30 seconds without neighing and taking off in the middle of the arena.

Common ways to deal with this issue might have been taking another horse into the arena with us or longeing the horse in the arena until he grew tired of neighing and "learned" that being alone was okay. In the worst case, the horse's emotional behavior might have been ignored by his trainer or a couple of strong men would be brought in to hold the horse and force him to comply with saddling. All of these measures have been designed from the perspective of humans, not horses and, yes, they can work—depending on the horse. *But* they would usually mean a case of learning in a state of overstimulation or fear. And, as we have already learned, the horse's vast memory misses nothing. Whenever something happens in his life that reminds him of this moment of fear or coercion when he as a highly intelligent being was deprived of the option of flight and independently resolving the situation, he will try anything to defend himself to prevent it from happening again. This might later become bucking under saddle whenever he is ridden away from a group or constant fidgeting upon entering the arena. This demonstates that this horse's lifelong nervousness isn't due to his genetics but to how he has been trained.

A better way to handle this situation is what I did with the young horse who couldn't cope with being in the indoor by himself: I repeatedly led him into the arena and then back to his stall. I kept taking him back to his comfort zone and then bringing him a little farther out of it. I only took him so far that I could read in his body language and his respiratory and heart rate that two more steps would lead to release of adrenaline. I allowed the horse to stay in control of the situation. He was challenged but not overwhelmed by gradually being brought a little out of his comfort zone, training session by training session, and therefore he was able to enter a state of responsiveness rather than defensiveness.

Just outside the comfort zone is where the magic happens, and that is where the horse needs to be. We can create a moment of stress, but only to the extent where the horse can cope with it next to me, with me, and remaining aware of me—and then we might do a circle and go back to the beginning. We must work creatively and use our human ability to plan to gradually bring the horse closer to the goal.

Licking and Chewing—A Sign of Stress

"Licking and chewing" is an equine gesture that we will definitely experience in the course of the training process. It provides us with information about a change in the autonomic nervous system (the control system that acts largely unconsciously and regulates bodily functions) at that moment. When at rest, the horse is controlled by one of the two divisions of the autonomic nervous system—the *parasympathetic* nervous system, which is responsible for metabolism, recovery, and for building up the body's reserves. If I take the horse away from his herdmates, a situation that he finds stressful and frightening, his nervous system will then switch to the second of the two divisions: the *sympathetic nervous system*. This enhances the horse's ability to perform by preparing the body for fight or flight. As already mentioned, the heart rate and blood pressure increase as a result, and the metabolism is stimulated. The horse's pupils and his bronchi dilate. Secretions from the sweat glands increase and salivation decreases. The horse's lips and mouth will feel dry.

If the stress that the horse is experiencing is immediately reduced rather than increased (because, for example, I have noticed his body language signals and stopped the training session at the right moment), the autonomic nervous system will be controlled by the parasympathetic division again. His heart rate and blood pressure decrease, and his pupils and bronchi constrict. Saliva production resumes again, which in turn instigates the "licking and chewing" gestures we are all familiar with.

Licking and chewing, therefore, indicates the horse has been exposed to stress and can now relax again. It shows that he is being controlled by the parasympathetic nervous system again and is entering a "recovery phase."

If I just walk back and forth from the stall to the arena to the stall and don't take the horse out of his comfort zone, I won't see this licking and chewing. The ideal is to always remain just shy of triggering licking and chewing, but still to elicit a learning situation—that is, to get 15 feet into the arena and turn around *before* the horse displays any licking and chewing. In the next session, I might go 20 feet into the arena, so I have brought the horse just out of his comfort zone, but not asked too much of him.

Of course, if we are not getting closer to the goal, then no learning is taking place. We therefore shouldn't be afraid of triggering licking and chewing, but we need to deal with it carefully when it is observed. As a trainer, I know that the horse has experienced or is experiencing stress when I see this behavior, so I will now try to stay slightly below that level as we continue our work together. The solution isn't to just keep walking back and forth for days or weeks on end, repeating the same lesson until it finally works. The solution is to guarantee measurable learning success in every session by being aware of what I'm doing; following a concept; ensuring the horse is healthy, alert, fit, and can correctly interpret my body language signals; and applying learning theories from the horse's point of view. This will help me achieve the best possible training success.

As we implemented our research based on 800 horses at the Lewitz Stud, one thing became very clear. When I work with EBEC and strictly follow the path of the EBEC Pyramid, all of the horses have different problems at different times, but they all have the same foundation instilled within a very short time. Applying EBEC might sometimes feel slow from a human point of view—for example, many of us might think, "I don't have time to raise and lower the halter 10 times." But my investigations show that you get to the desired end result much faster by not skipping individual steps. Above all, rushing the process will cause problems that can ultimately necessitate very lengthy, expensive retraining or correction due to the function of the horse's brain. In the worst cases, rushed horses never compete or have to be sold at prices far lower than would have been possible had they fulfilled their "genetic potential." Nervous-looking young horses, or even adult horses, dancing and sweating beneath their rider in the arena, are not an uncommon sight. They are inattentive, anxious, worried, difficult to saddle, or are overwhelmed when approached by too many people at once. Sport horses who are rushed are never able to learn as well as horses who are given a solid foundation for learning and who have been able to develop their full potential.

"Slow and steady wins the race" is open to interpretation. The individual horse determines what is fast or slow, and how fast or slow the trainer can go, from his own perspective. To help us figure this out, the ethograms give us measurable parameters that we can all use as a guide with a little practice (see p. 88 for more about the ethograms).

SUMMARY

EBEC Level 1

- We ensure that the horse's basic needs are met.
- The horse must be alert, receptive, and relaxed.
- We keep stress levels low because this influences how information is stored in the horse's brain.
- We notice all of the horse's body language signals. The transition to Level 2 of the EBEC Pyramid stops being possible as soon as the horse's state changes to one of uncontrollable fear or his ability to concentrate is reduced.

EBEC Pyramid Level 2

Interspecies Communication as a Foundation

COMMUNICATION BETWEEN ANIMALS

Communication—the exchange or transfer of information—between animals is organized into several distinct types and can take place in different ways.

Proprioceptive Communication: Auto communication where a living creature is both the sender and the recipient of a message. An example of this is echolocation, seen in whales, dolphins, and bats.

Interspecies Communication: This takes place between animals of different species—for example, parasitism, mimicry, symbioses, as well as aggressive and defensive behavior. Interspecies communication also includes dialogue between people and horses.

Intraspecies Communication: *Intra*species communication is communication among animals of the same species. It can only take place when all participants use the same code and apply the same rules. Knowledge of the code and rules can be an instinctive ability that is present from birth or an innate behavior that has been learned or trained.

Unidirectional Communication: This is intraspecies communication in just one direction, from sender to recipient. Bees dance to tell other bees the location of a food source. The message conveyed through dance is unidirectional because the dance does not trigger a symbolic reaction in other bees, but a practical one.

Symmetrical Communication: Symmetrical communication, which has the potential for dialogue, is the opposite of unidirectional communication. One example would be the behavior of dogs during the ritual of making contact.

The content of animal signals is often ambiguous and dependent on the context in which it takes place. For example, the position of the sun plays an important role for the information conveyed by bees about the distance and direction of a food source. The relative position of the interacting animals to each other or in the perceptual field can also be significant. So, distance to other animals of the same species, the food source, or the structure or the nest can influence the content of the message.

COMMUNICATION BETWEEN PEOPLE AND HORSES

Communication between people and horses is *interspecies communication*—communication with gestures between different species. It is very important to closely observe the horse's movements and gestures and to correctly decode the information that the horse gives us. We can only enter into a dialogue and control the horse's behavior in our interests if we know what his signals mean.

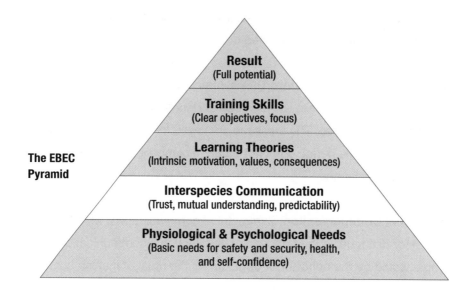

As horse people, we want our horses to be well and happy. Horses should also do what we want them to—behave calmly and dependably, be nice and friendly, and meet our needs. Then all is well with our and the horses' world.

This might seem rather egotistical at first, but that's okay given that we actually use horses for our purposes. I can support this view, because after all, I can make life good for my horses.

Horses have an excellent ability to learn, and they are adaptable, too. We have a lot to gain from realizing that horses remember almost everything that we teach them using classical or operant conditioning. We should therefore primarily concentrate on correctly decoding the horse's gestures so that we can respond to them meaningfully from the horse's point of view. As I've already mentioned, ethograms that apply to all horses, across all breeds, can help us with this (see p. 88).

In our attempts at communication, it is important that we don't imitate a horse's body language and try to assume the role of the lead horse or alpha animal. Neither should we imitate predators. We are human and nothing can change that. If we use our own instinctive gestures, then we will be natural and authentic. We make gestures and see from his behavior how the horse responds and decodes them.

We have to make sure that we don't react to his responses with punishments that correspond to our strategic thought processes. Examples of this would be assumptions such as: "If you don't stand still next to me, I'll send you out onto a circle until you give up and show some respect!" or "You're not standing still so I'm going to smack you!" Assumptions and actions of this kind come from the human perspective and aren't comprehensible to horses.

The first of these methods—sending away—copies behavior that can be observed in horses in the wild. However, when a human sends a horse away, the horse doesn't have the chance to actually escape the "attacking" human through flight. There is a high risk of stress as a result, and it can be so great that it blocks the horse's ability to learn for hours.

In the second example—hitting—the human expects the horse to obey the words and "stand still." The horse experiences on a physical and emotional level that the actions he has offered so far to the request—maybe walking forward or circling around the trainer—trigger aggressive behavior in the human. The reason for this may be that the horse has not yet learned the behavior "stand still next to the human" that the human expects. However, it could also be that the external circumstances have made the horse feel so unsure that he is too stressed to stand still. It is therefore essential to make factual observations and not respond by getting angry and acting aggressively.

If some external factor has caused the horse concern, then that could be why he doesn't stand still as soon as the human does. His internal system has probably sounded the alarm. If the horse had been able to recognize standing still as the correct behavioral response and been able to do it, he would have done it. Remember, horses do not have any strategic or planning capacity that would enable them to be deliberately disobedient or bad. External circumstances that prevent the horse from standing still could be simply an upturned bucket in the background that the horse cannot make out clearly so he can't understand what it is. We can help the horse by making the bucket stimulus recognizable for him.

Ethograms help us to understand whether the horse has caught sight of something worrying in front of or behind him. If we correctly interpret this from the horse's gestures, then it is helpful to bring this information into the equation (see p. 96).

For example, if the horse's ear position signals that there is something behind him, then we can walk in a little circle, following the angle of the horse's vision. As per his field of vision (see p. 101), moving in a circle will enable the horse to make out or decode the stimulus "bucket," and his concern will immediately subside. An aggressive response to not standing still, in the form of a smack or a shout, is never as effective as positive or negative reinforcement applied in accordance with learning theory. (We will discuss learning theories in Level 3 of the EBEC Pyramid—see p. 122.) The horse is not able to process the punishment. All we do is create uncertainty in him, and we have to accept the side effects of our actions in the form of additional stress. Our next lesson will be less focused and positive. Not because our horse is unwilling but because of the hormones circulating in him due to our actions.

ACOUSTIC COMMUNICATION IN HORSES

Our verbal language can be confusing for horses because they have no idea what our words mean. Horses can be conditioned to words, or certain sounds or tones, but we must always be aware that our strategic thoughts don't mean anything to them.

A horse won't decode "Stand still!" if he isn't able to stand still. Speech is consequently rather secondary in training horses. However, it can have beneficial side effects, for example, when the horse is fully focused on a stimulus that is worrying him. A shout or a sound from a different direction can distract him and change his focus. Horses are distractable animals and that can sometimes be useful.

When we give verbal expression to our happiness, our tone, a smile, or body language that radiates happiness can convey positive feelings to the horse. We just have to be aware that the words we speak are usually directly connected to our physical stimuli, such as stroking, kissing, or patting. However, there is so far no scientific proof that the horse "understands" that our words and our happiness are a response to the dressage test he just nailed.

A horse's acoustic language—the sounds that are audible to us—have been examined scientifically and consist of just a few different sounds. There are sounds that are produced via the horse's larynx and sounds that are produced without the larynx. Vocalizations are important information that is communicated to other herd members.

Acoustic signals are sometimes used in attack and defense rituals, but also to warn of danger or in friendly, playful exchange, as a greeting signal or for recognition. Three vocalizations are very important in EBEC training—the *blow* through the nostrils, the *snort*, and the *snore*. These three vocalizations give us important information about the inner state of the horse in training.

Blow

The *blow* is a non-pulsating broadband-level sound that results from a short, powerful expulsion of air from the nostrils. Blows can reach an audible range of 98 feet and are an expression of alarm. When a horse blows, it signals danger to all horses within that distance, perhaps, for example, because the horse concerned has discovered an unusual object a few yards away from him.

The average duration of a blow is half a second. It warns all the other horses in the immediate vicinity and informs them that something unknown has been discovered. It triggers an instinctive, unlearned alert in all other horses.

The nostrils dilate fully during this short blow, the mouth is closed, and a brief inability to move during and immediately after the sound is typical. Longer expulsions of air of between 0.6 and 1.3 seconds are often emitted afterward, and the olfactory investigation of the unfamiliar object will then follow.

When a horse who is being ridden in the indoor arena hears a blow from a horse who is being led past, outside the arena, he will immediately become nervous. His behavioral response will be directly controlled by his brain, without any of his own cognitive thought processes. He will immediately express his fear physically, maybe by lifting his head and speeding up. He will try to find out what is going on outside and whether there is any reason to take action in the form of fight or flight.

If we punish the horse in this scenario with a sharp tug on the reins, a smack with the whip, or other aggressive consequences, he will still do all he can to find out what is going on outside. He can't just imagine the worry isn't there and say, "Oh, sorry, I'm a little distracted at the moment. I totally forgot that we are in the arena and the doors are closed. Even if there is a bear outside, I should still pay attention to you! Sorry I messed up your lateral movement across the diagonal!"

Although they might seem logical from our human perspective, punitive measures in situations like these will not be effective, and the chances of achieving a positive training experience for the rest of the session after using them are low. Stress in horses is a combined psychological and biological behavioral response to either an old, stored experience that has already happened, or new, threatening circumstances.

The physiological behavioral response to stress is an extremely complex topic that has not yet been fully explored, but scientists agree that there are two types of stressors:

- First, *physical stressors*. These are physical injuries, strains, exhaustion, and changes in the environment.
- Second, *psychological stressors*. These include situations that cause the horse anxiety, worry, or fear. Science defines uncertainty and fear of the unknown as the two main psychological stressors for horses.

In the situation just described, the horse experiences the latter, a situation of psychological fear, triggered by the blow from the other horse. The horse expresses himself using interspecies communication. Those are the facts.

We as riders can use this scientific knowledge to stay calm and understand the situation from the horse's point of view. We can then behave in an empathetic and understanding way as we sit on the horse's back and consider how the horse is expressing himself physically. The positive effect of this is that the horse experiences support, instead of the rider adding even more stress to the already upsetting situation. If the horse's head and neck are being flexed with the rein contact, the rider could perhaps give forward slightly to make the horse feel that he is in control, and to prevent a conflict caused by the use of force.

When the horse gets too fast, it can be useful to divert his speed onto a circle. The horse is less likely to buck if he can still move forward, because he feels that he can escape the situation. He increases the distance to the blow with his own actions, which gives him a feeling of safety. Either way, the horse will only come back under control and start concentrating again when he can categorize the blow from the other horse.

This behavior is instinctive. The job of the rider is to remain understanding and redirect the horse's urge to move in as controlled a way as possible until, from the horse's point of view, the danger is no longer present. As soon as the horse's focus is back on the rider, work can resume. The rider will be able to feel this as soon as the horse accepts the rein aids again and bends and flexes as he normally does in focused work.

For the few seconds after the blow, the horse is able to perceive neither positive nor negative stimuli. The rider has to accept that she must abandon focused riding at this moment, since "sawing down" the horse's head, changing to a stronger bit, or using draw-reins or side-reins will not help. It is an unalterable reality that stress hormones in the bloodstream do not allow horses to be receptive to training or instructions.

Snort

The snort vocalization is important because it is often related to our actions. The snort is a broadband-level sound produced by forced exhalation through the nostrils and differs from the blow by an audible, fluttering pulsation of the nostrils that is visible to the rider. The horse's mouth remains closed during the snort. The average duration of a snort is 0.8 to 0.9 seconds, and it is audible up to a distance of 165 feet.

Snorts occur when the horse's nose is irritated, for example, by excessive dust in the arena, but also after high physical exertion or when the horse is restless and people are impeding his movement. This happens with excessive bending and flexing during ridden work or when the horse is contained behind a fence or barrier that he wants to but can't escape. Scientists emphasize that the snort is a "psychological displacement" whereby the horse attempts to express his unease. At this moment, he is unable or not allowed to give physical expression to his unease through movement. As trainers, we have to make sure that we notice the horse's snort when we cause him distress. If possible, we should release the horse from the situation so that he can relax and then give us his attention again.

If the horse snorts when he is being ridden or when he is alone in the field, Level 1 of the EBEC Pyramid is not being guaranteed. At that moment, we should notice that the horse has not understood an exercise or isn't able to cope with it. We should either remove the pressurizing stimulus caused by the rider or (as an example) take the horse turned out in the field back to his friends in the stable. In effect, we let the horse "get away with it," since he has limited capacity for learning in this situation. The same goes for us when a situation makes us nervous.

We must then use our ability to think strategically to create a training plan for the horse with the help of the EBEC Pyramid: In the next day's lesson, we will make sure that the horse's needs are fulfilled (Level 1—see p. 34). Then we check whether we are decoding the horse's gestures correctly. We think about how we can apply the learning theories and which rewarding or punitive stimuli will work best. And, in just a few training sessions, we can teach our horse how to behave while turned out by himself or how to deal with whatever triggered the snort when he was being ridden.

When you train the horse well, you will hear the snort less and less frequently. The older and better trained the horse is, and the better we have filled his "personal map" with positive experiences, the fewer conflicts will arise. Horses are extremely willing to learn and adapt.

Snore

The snore is another important signal that we should pay attention to in interspecies communication with horses. It is a broadband-level sound that is like a scratchy inhalation or an in-breath that just happens to be louder than usual. It lasts for around 1.0 to 1.8 seconds. We hear the snore under two circumstances: It is the preliminary sound, the advance warning of an alert blow (see p. 73) and can also be emitted when the horse is lying down.

The snore is important because, when we use EBEC, we concentrate on noticing reactions in the horse *before* situations escalate—that is, before horses spook or start bucking or rearing, and before stress hormones are released. This is what makes EBEC such an amazing tool; it is quick, safe, effective, and almost invisible. The sooner we recognize things that are unfamiliar or frightening to the horse, the more effectively we can immediately draw up a training plan that reinforces and then builds on the foundations of learning. We can enable the horse to develop his full potential.

Note, the aim is to correctly observe and classify the horse's gestures as early as possible, but *without judging them*. We want to take the right actions for the horse, before conflicts, problems, and misunderstandings escalate. This requires self-reflective action. If we notice a snore in training, we know that it is related to a moment of anxiety that we need to diffuse as quickly as possible. For example, we have to improve the horse's ability to see if there is something he doesn't recognize in the distance or in his blind spot. Or we may have applied the anxiety-causing stimulus ourselves and need to remove it. When this is the case, we may need to use the EBEC Pyramid to make the unfamiliar, frightening stimulus a familiar, accepted stimulus through appropriate training. If, however, we ignore the horse's behavior and the anxiety-producing stimulus appears again in the future, in the same or a different form, the horse's next behavioral response will come sooner and stronger. We want to avoid allowing such an issue to become established, because with good training, the horse should only be afraid of a few, totally unfamiliar, unpredictable things.

The chapter on applying the learning theories from the horse's perspective (see p. 122) explains how to trace the triggering stimulus that causes the horse to blow, snort, or snore, and then create a corresponding training plan to change how the horse behaves the next time a similar scenario is encountered.

PRINCIPLES OF TRAINING WITH EBEC

When we train with EBEC, we learn to create training plans that cover, as broadly as possible, the range of things that the horse has to learn, whatever his discipline.

At AKA we have drawn up a comprehensive catalog of skills that helps us and the horse feel calm and in control, because the horse learns the exercises he needs at the right times, and is then extremely well-prepared for whatever comes next.

A show jumper needs to be able to do different things than a dressage horse; a pleasure horse must have different skills than either of them. The focus for a pleasure horse would perhaps be more on remaining calm when ridden alone on roads and trails and accepting unfamiliar objects and scenes, while the competitive arena has very different and potentially frightening scenarios in store for a dressage horse. Meanwhile, the fences the course builders have deliberately tried to make as difficult as possible for the horse, in order to guarantee an exciting competition, pose other challenges. If a horse is prepared with the right set of skills, we minimize sources of errors and increase the horse's focus on the job in all these situations, a victory for us both.

That's the great thing about using a science-based training method. It can be applied in all situations and to all training goals and avoids producing problem horses, because it doesn't create potential for conflict.

NONVERBAL COMMUNICATION WITH HORSES

Nonverbal communication—communication with gestures—is important for an effective exchange of information with horses. It's a two-way street. Both sides, the horse and the person, send and receive gestures and signals and decode them. Some of this communication is *subconscious*, which we will discuss later (see p. 162). First let's look at the *conscious* interspecies communication process using body language. It is not only about reading and assessing gestures, but also movement in general.

Nonverbal communication means more than just paying attention and seeing what the horse does. We have to understand what his gestures say and mean. Everything is interrelated here, too, and must be looked at as a whole. Only paying attention to the movement of the horse's tail can be misleading if we don't also get information about the position of his ears, the expression of his eyes, or the placement of his legs.

Horses are animals of movement. They live in herds and rely almost entirely on perceiving body-language signals for exchanging information with other herd members, in order to make everyday life as conflict-free as possible. Horses only use vocalizations in specific situations. They are primarily silent, presumably to protect them from predators and to make them harder to find. But, they send signals that contain information all the time. Their movement or gestures can show intention to perform an action; provide information about current activity; communicate social status; show mood, emotions, and identity; and indicate the horse's psychological and emotional state. Nonverbal communication is two-way communication where one sends and another receives. The recipient's response to what is received creates a dialogue.

I have communicated with thousands of horses by copying their behavior. For example, I have used the "sending away" that can be observed in horses in the wild. I asked for activity when a behavior was unwanted and offered rest when the desired behavior was shown. I used this as a negative consequence for negative behavior, seeing it as an application that was a further development of the methods of the cavalry school, which tended to use punitive tools to change behavior.

When I began to scientifically investigate horse training methods, it became clear that while communicating with horses like this *can* work, it doesn't always. It all depends how much psychological stress the interaction causes the individual horse. Because stress has been scientifically proven to reduce the learning curve, I wanted to do better.

I took my first steps toward improving my methods at a small breeding farm in North Germany. I investigated which of my gestures instinctively triggered a certain behavior in the horse (these were not gestures where I copied a horse or imitated the lead mare, but human gestures. I am human and nothing can change that.) I went into the round pen with hundreds of horses, opened my hand, and looked directly into the horses' eyes to see whether this gesture would naturally cause them to move away from me. Sometimes it worked; sometimes it didn't. It depended on what I call the horse's internal "map"—what he had already learned in his life so far and on which information he had stored. When I opened my hand and looked straight at him, a horse might not move away from me; he might just look at me. But he might also run away at lightning speed. The key is therefore mutual encoding and decoding of gestures and signals.

How Do Different Horses React?

The process of interspecies communication with horses means observing and understanding which behavioral responses an individual horse gives to which of my gestures and signals—that is, how does he decode the gestures, from the very first moment I meet the horse, whether in the round pen, in the indoor arena, or in the stable.

The horse is right no matter how he behaves, because messages are created by how *recipients* decode them. A horse who, for example, has been longed a few times or maybe even longed his entire life in a halter, cavesson, or bit, is supposed to move away and walk in a circle around the trainer or he will be tapped or given a visual stimulus with the longe whip. (Note: I *never* recommend single-line longeing.) This horse, then, will respond with a flight response or by moving away when I go with him into the arena or round pen. This is a conditioned, trained behavior. If the horse is used to coming into an arena or round pen to practice standing still or staying with the trainer, he will have learned not to move away immediately. If in doubt, this horse will just keep standing in response to my open hand and direct gaze. He might look at me with a noticeable question in his eyes because he doesn't know what he is supposed to do. The most important thing when this occurs is to be clear about what I am asking of the horse. I have to be aware of this and use it to guide the rest of my actions and my training plan.

Basically, everything we do with horses is against their nature. Or rather, they are human ideas that horses would never come up with: Wearing a halter, standing tied, or walking into a trailer and not being able to get out, especially as a flight animal with no means of escape. Jumping fences in an indoor arena, "dancing" in a dressage test, jumping through flaming hoops in a horse agility course, or cantering through the countryside on a Saturday morning—none of this is part of a horse's natural behavior. Horses wouldn't choose activities of this kind because many of them are totally against their natural behavior patterns. From the horse's point of view, these activities are a needless waste of energy that take them away from the safety of the herd and could put their life at risk.

The horse's genetics can of course be part of the equation, but the training exercises that we use in order to bring about long-lasting behavioral changes are purely attributable to learning. We "call the tune" and we can each decide what we want horses to do for us. We can teach them anything we can think of.

Against or In the Nature of Horses?
No one type of riding is better, more beautiful, or healthier than another, and no one discipline deserves to be attacked by supporters of another. Everything should be looked at individually. Horses basically have to learn everything that we do with them. They have to understand that we have thought up these exercises for them.

They live with us in the environment we put them in. We determine their life, their future, and their existence. Whether the way they are kept and used is in or against the nature of horses, must be examined in each individual case. Scientific thinking is always based on fact. If the facts say that a horse is suffering, we need to make changes in the interests of his welfare. Obviously not every competition horse is suffering, just as not every field-kept horse is content. If we consider this question from the horse's point of view, things will be better for the horse because he has a better chance of responding to our objectives.

THE HORSE'S RESPONSES TO COMMUNICATION

The first time a horse is ridden in an indoor arena, he has no idea what he is supposed to do. If in doubt, he will just run. If he has already learned that whenever he comes into the arena, he is supposed to stand in the middle for some time while people put a saddle on him, and then walk for five minutes to the right, then the horse will rely on this experience to know what to expect. This is not a cognitive decision, but knowledge based on a previous experience. Every behavior offered comes from practice. If an action is positively reinforced, the horse will offer it again. However, the horse himself does not know or is not capable of thinking about what is being asked of him.

If the horse has previously come into the arena without a rider, and for example, been driven straight into free jumping or just turned loose so that he can roll, there is a high chance that he will offer these behaviors again. We have given him an option for behavior, so he will offer it again should the occasion arise. The problem is, this can result in the horse rolling in the indoor at what we consider to be the wrong moment—when we have put on his saddle, for example. Horses are not able to think this through. When in doubt, he will roll with the saddle on. If, however, he is given a signal that he can decode and that clearly tells him that he can only roll when his lead rope has been unclipped from his halter and he is loose, without tack, he will only do it then.

Remember, horses work according to the following formula: "What did I do and how did it work out for me?" If it went well for him, he will offer this behavior again. If it didn't go so well for him, his limbic system will do everything it can to get him out of the situation quickly or not get into it in the first place, as soon as the first stimulus gives him an inkling that this could be related to an earlier situation. Depending on how we want to work with the horse, there are effective gestures and signals that the horse can successfully associate with certain actions and correspondingly decode.

THE HORSE'S EXPRESSIVE BEHAVIOR

Let's say I have a young, untrained horse who I am supposed to start under saddle. The objectives are for him to become familiar with his first saddle, first rider, first bridle, and first long-reining sessions. First, I think very carefully about which gestures and signals the horse is supposed to decode, and I show him this information in the very first training session. I determine these gestures and their meaning because they are a result of conditioning. Whatever happens, I want to keep the stress level low. The horse should not be stressed, and he should give the correct behavioral response. Perhaps I decide that it would be very useful to me if the horse moved away from me in response to direct eye contact. In my experience, horses find this easy, presumably because it provokes an association with communication among horses.

Intraspecies communication between horses is extremely effective. From a scientific point of view, we differentiate between *whole-body communication* and communication *using parts of the body*—gesturing with the head or legs, for example. However, it should also be said that head and leg gestures normally go hand in hand with changing ear, tail, or poll positions and facial expressions.

We decide how the horse decodes our gestures, but the science-based ethograms I've included in this chapter enable us to reliably and correctly classify every individual gesture of the horse's body and respond to it.

People are not always consistent in their body language communication when handling horses. This makes it extremely difficult for horses to correctly assess situations and to offer the behavior we want. You could also say they "guess," which is a human concept but maybe quite a good way of describing what the horse does to interpret our signals.

Sometimes we look horses straight in the eye with the intention of getting them to move away from us, when working in the round pen for example. Other times we look them straight in the eye to discipline them when they misbehave and to get them to pay more attention.

We cause similar confusion with rewarding or punitive stimuli, such as patting the neck. We will talk more about this later when we look more closely at learning theories in Level 3 of the EBEC Pyramid (see p. 122).

Again consider the horse I was supposed to be starting. He will have to learn a lot; everything is new and he won't be able to do anything I want unless I am as calm, clear, stress-free, and predictable as possible. That means not causing any stress or increasing his heart rate, as we discussed in the last chapter. Let's imagine the horse is in the round pen, and I use gestures that are practical, valuable, and easy for the horse to understand because they are similar to his own means of intraspecies communication. I think through my objective. I need to keep looking at the big picture and the situation as a whole, in order to recognize and estimate the consequences that the information conveyed could have for the horse.

The horse should stand calmly in the middle when asked because that is where I will want him to be when I put the saddle on in the next training session. The more calmly he stands in the middle of the pen, the better chance girthing will go smoothly, and the higher the chance that the horse will be successful at promptly offering the correct behavior throughout the tacking up process—that is, he'll be more likely to "get it right" the first time. I can then praise him and positively reinforce the behavior. (More about that in Level 3 of the EBEC Pyramid—see p. 122.) Now I use the gestures "eye to eye" (looking at him directly) and "open hand" (raising my open hand) to move the horse away from me in the pen. It's very important that I don't chase the horse. He can move off in walk, but he can also choose a faster gait. I can leave that open to him at this moment. The main thing is for there to be forward movement. I move away from the horse at a 45-degree angle, allowing him to increase the distance between us, and keep looking into his eyes.

Horses are highly intelligent and social beings who will look for any human gesture that they can assess and decode. I know the horse may offer various behaviors. While I am looking at him, the horse is allowed to do anything except come to me.

I intensify the driving stimulus. An open hand, the front of my shoulder turned toward the horse—these gestures cause the horse to move forward as I want him to. There are gestures and signals from interspecies communication with other species to which the horse will instinctively respond with behavior that I consider to be correct. These include all predator gestures, but I can only intensify use of these gestures to the extent that I remain below the stress-production threshold. This is easy to montior in horses if I am familiar with the ethograms (see p. 88).

The horse decodes my gestures and signals and offers a behavior. He moves away because the front of my shoulder is toward him and because my eyes are fixed on his. I then turn him and he changes direction. I let him turn back and then move my shoulder away and encourage him to come to me. I walk in arcs so that I am visible to the horse throughout his entire field of vision (see p. 101).

When the horse comes to me, I give him a positive reinforcer that he is receptive to. This could be a rub between his eyes or a stroke on his neck. When I take the horse back into the arena the next day and repeat my actions, he will offer everything that he experienced without stress the previous day, as a result of conditioning, and the positive experience will be repeated. He decodes my gestures and signals because I have enabled him to do so. He decodes them in the way that I want him to because I have been clear. We have increasingly positive experiences together. Cooperating with me is something he considers pleasant, which creates a good foundation for communication going forward.

I need to stay true to the gesture I am using in our communication. I have to stick to the pattern. The horse should always remain still when I am standing level with his shoulder or at his head. When I am in the middle of the arena; when I move from the horse's right side to his left side; when saddling up; when attaching the long reins; when mounting up, grooming, being examined by the vet—the horse should stand still for all of these activities. Then, when I step back from his head at a 45-degree angle, he should move away.

Establishing these rules early on are of enormous benefit for the whole of the horse's life. No matter who he is sold to, he will offer this behavior in all situations, unless he has a confusing or negative experience. This horse will offer his learned behavior and it will almost always be right. When I stand, he should stand, too. When he is asked to move out on the track, he should go forward. He becomes equipped with this specific, clear, and very significant decoding ability in the interspecies communication process. The horse can always expect the same consequences.

Horses tend to conserve energy and want to live a calm and predictable life, so this kind of training is exceptionally well suited to their needs. We can integrate even more gestures than I have described here. How we teach them will become clear when we get to Level 3 of the EBEC Pyramid (see p. 122). We know what we want and how the horse decodes *our* gestures, but working with horses is about two-way communication. It now gets a little more complicated and takes a little more practice, because horses have a number of ways of expressing themselves and making themselves understood.

ETHOGRAMS: THE VOCABULARY BOOK OF EBEC TRAINING

Communication using body language signals is extremely important for social interaction in the herd, but also for exchanging information with people. The primary signals for exchanging information can be *visual, acoustic, tactile,* or *chemical*.

Social interactions usually involve more than just one element. The intensity of the horse's expression has been proven to be dependent on the intensity of the stimulation—that is, the intensity of the trigger of a behavior. We can observe this, for example, when one horse wants another horse to allow him to eat at the hay pile. When the hungry horse is supposed to move away from the hay pile, the horse who is trying to send him away begins with a firm "eye-to-eye" contact, and adds laid-back ears and positioning of the shoulder. He then approaches the recipient horse and intensifies his expression. If the hungry horse ignores the gestures, the other horse will successively intensify his expression.

Communication is therefore dependent on social interaction, but also on the situation. Is it a matter of survival, is danger imminent, or is it just about establishing an order for drinking at the water trough?

In an interaction with a human, this might look like a horse with a full belly nuzzling gently at a pocket where he is used to finding a treat. However, if the horse is hungry and really wants something to eat because the amount he instinctively needs to chew per minute to have a healthy metabolism has not been satisfied, then he might shove at the human's pocket or maybe even bite, and he won't give up. In training, it might be the case that the horse I want to free-longe takes off in canter rather than staying in walk if there are too many things going on around him that unsettle him, such as a tractor grading the outdoor arena or the sound of other horses being worked hard in the indoor arena.

The function of physical expressive behavior using body language gestures is attributed firstly to signals from the environment that affect the horse, and secondly, to signals that affect the reactions between the sender and recipient. The gestures of interspecies communication are therefore dependent on both the environment and on the contact between the communication partners.

How Do I Read the Ethograms?

An ethogram defines all observable and clearly distinguishable behaviors of a horse, in writing or using graphics.

The equine ethograms that we need in order to be able to correctly apply EBEC to the horse and that are shown in this book, show the gestures of the head and tail. (Note: The ethograms shared here are only part of what we have developed at AKA. For more information and the complete ethograms from my program, see p. 169.) You should focus on the changing behavior of the eyes, ears, and nostrils, and the position of the head and tail. Movements of the muscles and legs are also important; however, they require a trained eye and practical instructions that help you become able to better assess the horse as a whole.

There is also a "neutral state" that describes horse behavior in a stress-free situation. A stimulus causes behavior to change in stages. The gestures that are described by a behavior normally follow each other. Since horses are able to react extremely quickly to external stimuli, it is necessary to increase the intensity of the stimulus in such a way that changes can be visible to the human eye and can therefore be influenced. Differentiating body language as a behavioral response to stimuli from the front, side, or from behind is very important in training and communicating with horses. Looking at the ethograms enables us to get an early read on the smallest changes that point to the horse's next movement and gesture.

These ethograms are based on scientific research. We can count on the fact that every healthy horse will act as shown in them, so the ethograms enable us to exert our influence early. We don't have to wait for the horse to rear or buck to react, which is what trainers have long done. Problem horses are produced when we wait until *after* a behavior is happening to try and correct it. Instead, we can now redirect the unwanted behavior using knowledge of behavior from horse to horse, in a way that is more comprehensible to the horse and that allows us to correct the bad habit as best we can. We now have a scientifically sound method that prevents rather than creates problem behavior.

Now let's spend a little time studying the gestures that ideally enable us to recognize what stimulus the horse has perceived at a precise moment. This is valuable, visible information for us humans: It enables us to be much safer with horses, because the level of intensity in the horse and also within us can be minimized. Horses are also better able to learn and can better develop their full potential when the release of stress hormones is minimized.

A key to this is knowing at the right moment where the horse's focus is directed and where the horse is on the pressure scale. We can then very calmly introduce a movement that immediately returns his behavior to normal and doesn't allow it to escalate. We often don't know which stimulus the horse is perceiving because, from our perspective, we can logically recognize everything. Again we will find that the art is being able to adopt the horse's perspective. Knowing whether it's the bucket or the wheelbarrow or the colored pole or the flowerpot that has been added to the course as decoration that is causing his behavior will significantly help in your interspecies communication. It is a definite plus, not just for the horse, but also for me as a trainer, if my full attention is on what my horse is perceiving from his perspective when I ride into a show jumping ring, dressage arena, down an unfamiliar trail in the woods, or even just when walking down the aisle between the stalls or leading the horse out to the field. Together we will become less fearful, more content in each other's company, and more successful.

Gestures of Full Body Expression

Horses can give us a lot of information with just the physical expression presented when standing, without other head movements or facial expressions being necessary. They give us information about their emotional and physical state, as well as their overall health. We can get comprehensive information by looking at the placement of the legs and feet, as well as the carriage of the head, neck, and tail.

This book doesn't deal with health-related postures or misalignment. We are focusing on the interspecies communication that is important for training and communicating with a healthy horse (where the First Level of the EBEC Pyramid has been fulfilled—see p. 36). The horse we are working with is presumed receptive and healthy.

Gestures with the Legs and Hooves

Movements of the horse's legs are important for us because they give us information about social interaction. When we are communicating with the horse we are in a social situation together. This is two-way communication from human to horse and from horse to human.

It is also important to distinguish between different leg movements in horses, because some of them are threatening gestures.

Kicking (A) is done with one or both hindlegs, while the forelegs remain in contact with the ground. It is an aggressive stance. The horse shifts his weight onto his forelegs, quickly flexes and raises his hindquarters, and rapidly strikes out behind. The neck is usually lowered and stretched forward during this process.

"Knocking," stamping, or lifting the hind leg (B) are related actions. The hind leg lift is often observed when dams want to prevent their foals from getting to their udders. All three movements can be very intense.

Striking (C) is a rapid movement with one or both forelegs toward another individual. It is a blow or a threat. One foreleg is often seen to initially remain in contact with the ground.

Pawing (D) is similar to striking with one leg, but it is considerably slower

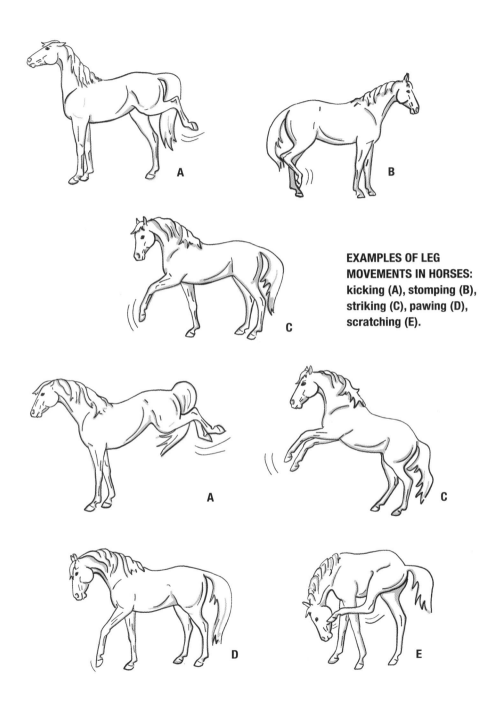

EXAMPLES OF LEG MOVEMENTS IN HORSES: kicking (A), stomping (B), striking (C), pawing (D), scratching (E).

and looks more like a scraping movement. It is also used to investigate a surface or an object.

Scratching (E) is done with a hind leg and is mainly observed in young horses.

Leg movements such as kicking and striking are some of the horse's agonistic behaviors that appear in conflicts with adversaries. Legs can be raised quickly; they are very expressive and can be read as preparation for attack or as a threat signal. When horses kick, they rarely intend to actually make physical contact with the object concerned. Kicking should be classed as more of a warning signal to give the horse some space and is either an offensive or defensive warning to get the other party to back off.

"Knocking" and stamping with the legs are also agonistic behaviors. The contact with the ground intensifies the gesture because the horse adds an audible stimulus. For example, horses eating in a herd are seen to noisily stamp their front or back feet as soon as another horse approaches. It can be classified as a kind of protest signal. The horse who is eating won't just forcefully kick out with the ill-willed intention of injuring the other horse. The hind leg lift is an advance warning of a kick that might follow. It is the *intention* of a kick.

If the horse also puts his ears flat back at the same time as giving a firm, loud stamp or pawing—for example, when being saddled or mounted—the horse is signaling a protest. He is making a complaint, so to speak. Something is bothering him. We need to pay attention to this. Instead of just shouting at the horse or smacking him, we should recognize what the triggering stimulus is and what could be causing the protest. After all, we know that horses don't "play" at protest, attack, or defense (see the horse's common "play" movements on p. 95). From their point of view, there is a valid reason for their behavior, and we need to find out what it is.

EXAMPLES OF PLAY BEHAVIOR IN HORSES: A foal plays through movement (top three horses), and an adult plays with an object (bottom four horses).

It could be that the horse has had bad experiences with being saddled—the nerve endings in the girth area might be damaged, or there might be something in the environment where the horse is tied up that he can't classify or can't see because the movement of his head is restricted. It is something that he feels cautious or skeptical about. If the object suddenly moves, such as a bucket that is knocked over, the horse can get such a fright that he jumps a foot in the air from a standstill and injures himself. This is why we need to pay attention to *all* of the horse's gestures.

Pawing usually occurs in situations of conflict. However, it is also seen when a horse is preparing to roll, investigating unsafe ground, digging for food, or investigating a food source. Raising alternate front hooves can be a sign of uneasiness and general unrest. A nervous horse approaches an unfamiliar stimulus and paws it and then jerks back away from it as an alarm signal for other horses. You may have seen this behavior before when a rider tries to grab a cooler hanging on the arena fence when a schooling session is done or when she tries to retrieve the whip that hangs on the wall. The horse approaches cautiously, tenses up, and makes a pawing, investigative, and slightly see-sawing movement. If he jerks back and makes noise when he does so, his behavior serves as a warning for the other horses. I have often found in this situation that all the other horses in the arena spook at the same time. The horse is providing a visual but also acoustic warning signal that triggers an instinctive, innate flight behavior in any horses nearby. If, however, his investigative movement is not sudden or fearful, it signals to all the other horses that the stimulus is nothing to worry about.

When we know all these warning signs then we can reliably prevent situations of alarm. This puts the trainer in a position where she can actively divert the relevant warning signs into other movements so that the next stage in the escalation of behavior isn't reached.

Impressive Gestures

A high, arched neck carriage with the nose tucked in, pricked ears, and a raised tail are gestures used by stallions in sexual displays. Some advanced dressage movements, such as piaffe, aim to imitate these gestures. Other horses in the arena can become extremely nervous when a horse approaches them in piaffe with this powerful physical expression. If a wild horse approaches another horse in this way, such as a stallion approaching a mare during courtship, then the mare will turn around and kick out or take flight if she isn't ready to mate. That isn't possible when being schooled with other horses and riders in an arena.

This is why it is important to pay close attention to all gestures, because they are warning signs for subsequent behaviors. A horse who might be approaching the piaffing horse on the inside track doesn't have the capacity to think that the gesture actually doesn't apply to him, and that it is a desirable, athletic exercise for a sport horse, one we humans have invented and find attractive.

While a highly trained dressage horse might have a fairly calm appearance in piaffe, when the piaffe isn't done well enough and the horse's neck is overbent, perhaps with his head held by a strong bit while the rider asks for the required stepping movement in the hindquarters with a powerful driving aid, and the horse is snorting loudly because his larynx is under pressure, then it might be a very intimidating expression for other horses in the vicinity. It is not uncommon for other horses in the arena to take off at this sight or stop wanting to pass by the horse.

The answer lies in intraspecies communication—the exchange of information between horses. Information is obviously exchanged between horses when we are riding. Horses use *inter*species communication to try to decode *our* gestures in order to understand what to do, but at the same time, *intra*species communication also takes place with any other horses in their field of vision. Yes, we can condition horses using learning theories not to react to those communications. But we are well advised to think not from our perspective, but from the horse's point of view in these scenarios. If your horse evades the rein aids because he is trying to find out whether or not the horse who is trotting toward him is displaying visible sexual gestures, how you react will influence his possible responses in similar situations in the future. If you don't allow his evasion, use force to bend him, or even give him a poke with your spurs or a smack with the whip, the horse might associate any of these painful stimuli with being approached by the other horse. As a consequence, his amygdala will release stress messengers in the hippocampus next time around, in order to prevent the other horse from getting as close to him. You have created a frightened horse who perhaps now can no longer be ridden indoors with other horses. He will be alert for information from the trotting sounds of the other horses, the tactile feel of vibrations through the ground, and if possible, visual signals to warn him of a similar scenario and spook early on. In the worst case, the horse's brain will then be devoted to defensive and protective behavior, rather than to your actual training exercises, for the entire riding session. Level 1 of the EBEC Pyramid is no longer in place.

However, if we are aware of this and understand the horse's gestures, a single circle at the right moment, at my horse's first physical sign of unrest, can cause the entire training session to take a different course.

Facial Expressions

Changes in equine facial expression are very easy to notice. If we know what is normal for a horse at rest, we will notice his other possible forms of expression. There are many gestures that we can pay attention to. The movements of the lips, tension in the muscle tone of the face, tension or relaxation in the corners of the mouth, and changes in the outlines of the nostrils—whether or not they are wrinkled.

I look at the horse's face as soon as I go into his stall and greet him—and make note of his expression. I always take a few minutes to stand quietly, listen to the horse's breathing, and touch him. I allow the horse to notice and feel my gentleness, and then I won't deviate from this inner, compassionate composure during the training session to follow. And if for some reason I can't manage that, I pull back. I try to keep any annoyance or anger with myself for not getting something right or any cluelessness that manifests itself through lack of skill, away from the horse, because those are *my* problems, not the horse's problems. I don't leave this humane perspective at any point. With this quick pre-training ritual, I nonverbally give the horse the opportunity to get to know me and to perceive my inner calm. I gently stroke his soft neck with the flat of my hand, I move his forelock out of his eyes, maybe give him a little kiss and let him rest his head on my shoulder, where he can smell, feel, and see my pulse, my scent, my compassion, and my relaxation.

There is no scientific proof that this is beneficial for the training session from the horse's point of view, but it gives me the chance to experience the horse in a neutral state and to consequently be able to recognize more easily when the horse changes mentally, using the ethograms we are discussing. We are both in a neutral state in this moment. I really allow the interspecies communication to develop and attach great importance to the ritual. It's not a long cuddling session, but a kind of 60-second check-in. In technical terms, it's more about the chance to mutually encode and decode a few gestures and signals, and I recognize this horse's neutral state in his familiar surroundings and look out for any changes as we work together.

When studying the horse, I pay most attention to the eyes, the corners

GESTURES WHEN RIDDEN AND THEIR MEANINGS

Ears forward: harmony of the aids, willingness to cooperate, attention forward.

Ears to the sides: attention to the sides.

Ears back: fear, submission, unease, and readiness to take defensive action, pain, attention backward.

Ears flat back: aggression, massive defensive reaction to the rider's influence, extreme fear.

Mouth closed with elongated nostrils: attentiveness, fear.

Mouth clamped shut with elongated, flared nostrils: fear.

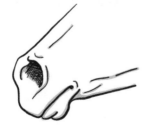

Flared nostrils: exertion, arousal, fear.

Constricted nostrils with wrinkled nose: pain, uneasiness.

of the mouth, the nostrils, and the ears. Using this book, I hope to train people of all levels how to perceive subtle details of the horse's expression so that they no longer have to try to analyze everything, but can instead recognize and classify a sign of a movement before it becomes a movement.

Attention Is Directed Forward

The ethogram of the facial expression when the horse's attention is directed forward, shows the features of the horse's expression when he is perceiving something in front of him by looking at it, listening to it, smelling it, and sometimes also touching it.

The horse pricks his ears and rotates them so that they are pointing forward. The horse is looking forward and his binocular field of vision allows him to see almost all around him. He adjusts the angle of his head and poll to enable the sensory receptors to be activated. The horse raises his poll and tucks in his head to allow him to inspect objects that are farther away. His nostrils are moderately flared, especially when he is sniffing or can smell something a greater distance away.

Horses behave differently when they perceive something to the side or something from the side catches their attention.

Attention Is Directed to the Side

The facial expression during lateral perception is very important for interspecies communication. It is characterized by general relaxation, where the eyes are normally looking to the side and the ears are also slightly to the side. Unless they are turned toward something coming from one side, the horse's sensory receptors aren't focused on anything specific, but are totally relaxed.

Once they realize I'm not up to anything, horses often direct their attention to the side during my 60-second stall check-in. They usually do not

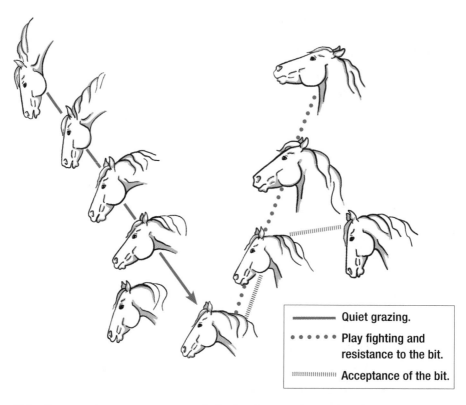

———— Quiet grazing.
• • • • • Play fighting and resistance to the bit.
ııııııııııııı Acceptance of the bit.

This ethogram shows some common behavioral expressions when the horse's attention is directed to the side. Note that only a few specific behaviors are labeled. For the complete ethogram, see p. 169.

show any real concern because there is no reason to be particularly alert.

We can see these sideways-directed gestures in horses both when they are standing and in motion. They are facial expressions that are also often seen in well-ridden horses when they are relaxed, in predictable routine work, or in a process that is well-known to them and gives them no cause for concern. Horses in this situation are particularly receptive and I feel able to try out a couple of new things under saddle. It's good to be able to return to this sideways-directed facial expression, because it shows that the horse is relaxing into his routine again. It is often observed in horses at competition stables during routine braiding, when the horses have worked and are contented, or in backyard barns when they know exactly what their owner's next step is. These relaxed, "chilled out" horses can be found in all disciplines.

Attention Is Directed to the Rear

Rear perception is characterized by the horse's eyes and ears being directed backward—as far as possible. The eyes and ears rotate backward

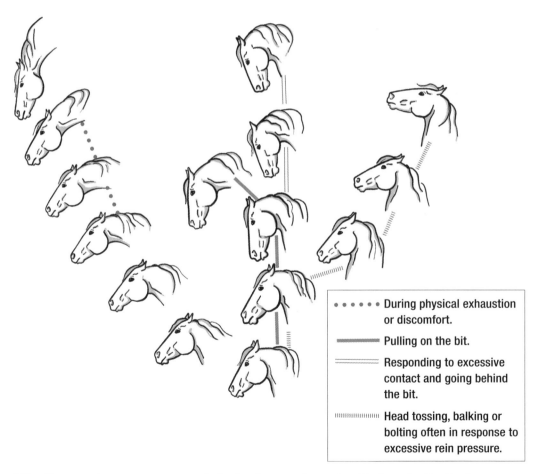

- • • • • During physical exhaustion or discomfort.
- ——— Pulling on the bit.
- ═══ Responding to excessive contact and going behind the bit.
- ⁞⁞⁞⁞⁞ Head tossing, balking or bolting often in response to excessive rein pressure.

This ethogram shows some common behavioral expressions when the horse's attention is directed to the rear. Note that only a few specific behaviors are labeled. For the complete ethogram, see p. 169.

FROM THE HORSE'S POINT OF VIEW • 103

and the horse's mouth is normally closed, except when he is neighing to give an alarm call or when his rider is trying to get control by pulling on the bit in his mouth. These physical expressions are not only seen when the horse is processing visual or acoustic stimuli, but also when the horse is stressed, feeling unwell, or anxious about his rider.

Rear perception is very important to us, especially when we are riding the horse with contact on the reins, long-lining or double longeing, or when the horse is standing in cross-ties in the stable. Many accidents can be prevented in these situations where everything we do happens behind the horse in the blind spot around his hindquarters where he can't see.

Remember, horses can see almost all the way around them (see p. 101). They see everything on their right side with their right eye only, and everything on their left side with their left eye only, and only around 20 percent of information is transferred from one hemisphere of the brain to the other.

If we restrain the horse's head and leave him unable to classify a potential stimulus from behind, it can result in numerous defensive behaviors, including those that can end fatally with him falling over backward. If the horse is free to do so, he will try to turn around, either to get away from the stimulus or to try to get it in his binocular vision and therefore direct his attention forward.

We must always pay attention when the horse's facial expression shows that his attention is on something behind him, even when someone else is riding. We need to be brave enough to interrupt the rider and stop the scenario because we know that stress hormones are beginning to be released in the horse. Adopting the "don't let him get away with it" approach isn't advisable when we see these expressions. It is more helpful to take an attitude of, "Oops, I didn't express myself well there. Sorry, I'll release the pressure, start again, and try to communicate more clearly." The latter attitude prevents you from creating an anxious problem horse. Scientific study shows us that if a rider keeps going and increases the pressure in a situation where the horse is processing one or more stimuli, the increased level of "alert" is noticeable in the horse's whole body.

I always sympathize when I see horses being taught to piaffe in a way that frightens them. This happens when the rider either sits on the horse or drives him from the ground, holding the horse in at the front with a curb bit or cavesson so that his field of vision is extremely limited. At this point, the horse can often only see the arena surface on the ground in front of his feet and can hardly make out anything else. If the trainer now hits him with the whip precisely in his blind spot to drive on the movement of the hindquarters from behind, the horse's facial expression will show that his attention is directed backward. We sometimes also see horses kicking out behind when they get so scared that they decide to kick out at nothing, into their blind spot, just in case the "attacker" is a predator—not a whip.

There are examples from other equestrian disciplines, as well. We often witness rear-directed perception in free jumping when the horse initially comes into the arena with his attention forward, but then people raise one of the poles just as he's jumping ("poling" or "rapping") to suddenly make the jump higher than it really is so that the horse jumps higher the next time around. Making the horse feel he has "misjudged" the height like this comes at the expense of his self-confidence and ability to perform and can cause lasting insecurity. Rear-focused perception and the associated fear also come in when the horse is prompted to clear a large jump with a longe whip behind him. You don't actually need to bother with subsequent repetitions in this case as, full of fear, the horse will now just fly over the jumps at full speed. This has been scientifically proven to negatively affect the learning curve and can cause anything from a long-term fear of poles and jumps to irreparable damage.

Techniques like this might seem like a short-term solution for a quick sale, but they are harmful in the long run. The horse will never be able to reach his full potential, the highest level of the EBEC Pyramid (see chapter 7, p. 162).

Ignoring backward-directed perception can also have a negative effect on the performance of racehorses in the starting gate.

The consequences of human ignorance can be observed in all disciplines, competitive or pleasure. It is therefore essential that we recognize the horse's gradual change in perception early on and note: "I'd better do a turn, a circle, or remove the stimulus." This can be a huge relief from the horse's point of view.

EBEC works by applying the learning theories to introduce a positive or negative consequence for a behavior *as soon as* the horse shows or suggests it. The consequence doesn't just come once the behavioral response is so strong that it is very difficult to change, because then the behavior has already been completed and, in the horse's eyes, belongs in the past. This takes a little practice and is a skill that we can acquire and which I teach in my AKA courses. It is good at first to have an instructor familiar with EBEC who can give you appropriate instructions at exactly the right moment. Timing is key with horses. This becomes clear when we look at the ethogram for expressions of alarm (see p. 107). If we don't pay attention and don't decode the horse's interspecies communication, the horse's behavior can quite quickly become aggressive, as clarified by the ethogram on page 109. Once the horse has become aggressive, it is difficult to find an understanding or achieve training objectives.

Expressions of Alarm

The expression of alarm in horses is shown by wide eyes, flickering ears, a tense mouth, and flared nostrils. The alarm messengers and hormones that are sent out into the blood circulation transform the horse's previously relaxed body into visible full-body tension that is often accompanied by sudden twitching movements as the body instinctively recoils from the stimulus. This is sometimes a hunched-back, scrambling motion. Sweat production starts very quickly, and there is a clear increase in the horse's

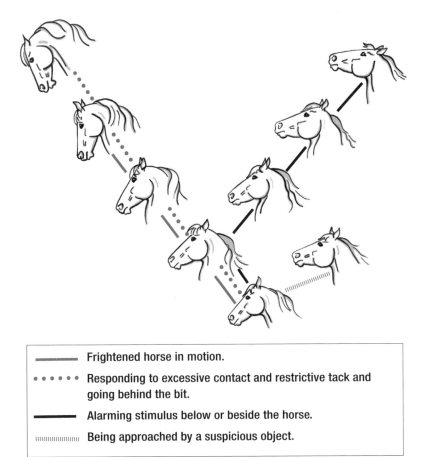

——— Frightened horse in motion.
• • • • • Responding to excessive contact and restrictive tack and going behind the bit.
——— Alarming stimulus below or beside the horse.
ııııııııı Being approached by a suspicious object.

This ethogram shows some common behavioral expressions of alarm. Note that only a few specific behaviors are labeled. For the complete ethogram, see p. 169.

pulse and respiratory rate. The horse's inner mental state varies from suspicious to high alert to extreme fear.

Expressions of Aggression

The expression of aggression can be recognized by the ears pinned flat back against the skull. The horse's eyes will be wide open, and they will normally be looking in the direction of the object that is triggering the behavior. The nostrils will be flared and drawn back, causing wrinkles on the top.

A high level of physical tension has also been proven to occur in the expression of aggression. If he is able to, the horse will have his mouth slightly open. In an extreme expression, the incisors will be displayed to convey a bite or the threat of a bite. Aggressive body language has many variations in the neck, poll, and head carriage that depend slightly on whether it is taking place between horses or against other potential threats or attackers, regardless of whether they are human or animal.

Suppression of Expressive Behavior

Many tools in the horse industry have been developed to suppress the expressive behavior of horses. For example, it can be difficult to clearly make out the position of the corners of the mouth when the noseband or bit is not done up or positioned in a humane and horse-friendly way. Nosebands and chinstraps are often too tight, which prevents the horse from being able to move his mouth when being ridden or stretch out his tongue.

Side-reins are used when the horse won't relax his back and flex his poll to the reins of his own accord. Spurs are used when the horse doesn't understand the forward leg aid. If the horse doesn't "understand" these spurs, then sharper spurs are used.

Numerous types of whips are also used when interspecies

////////	Biting or bite threats.
- - - -	Driving or dispersing others, sometimes with snake-like head swinging.
• • • • •	Vigorous approach of a stallion toward another male horse.
‖‖‖‖‖‖	Aggressive displays used in inter-male fighting; also a prelude to bucking in a ridden horse.
◇◇◇◇◇	Kicking or kick threats.
———	Used with foreleg striking, rearing, kicking, pushing, and avoidance; also a fear response in a head-shy horse.

This ethogram shows some common behavioral expressions of aggression. Note that only a few specific behaviors are labeled. For the complete ethogram, see p. 169.

communication has not been predictable enough for the horse, for example, when the horse doesn't offer forward movement on the ground in response to the signal with the trainer's shoulder or her open hand.

In many cases, tools like these don't solve the problem and often make it worse. If you use them, it is important to pay attention to getting the timing right and removing their influence appropriately. Otherwise you are training the horse to go forward in response to a tap or smack with the whip instead of the leg aid. If I am more aware of behaviors and gestures, then I can make the best decision for my horse and me, because I determine which behavioral response my horse should present in response to which stimulus.

Please bear in mind that horses don't do any of the activities we do with them on their own, of their own accord. We have thought up virtually all of the training exercises that horses are expected to do, apart from intra- and interspecies communication and the horse's instinctive behaviors. Horses wouldn't jump a course of show jumps by themselves without any stimulus. Horses wouldn't meet up by the oak tree at three o'clock to run a race or trot around an arena in a dressage competition. Neither would they choose to go for a walk in the woods alone, away from the safety of their herd.

The horse must be taught all of these requirements through Level 3 of the EBEC Pyramid (see p. 125). We base our training process on a conditioned stimulus-response behavior that is trained—that is, learned and based on repetition.

As trainers, we need to be aware of the consequences for the horse's learning when we decide to use certain additional pieces of equipment, for example, spurs as the driving aid or stimulus. A good tip is to take a quick look at the "big picture," to look in from an outsider's perspective, and then consider whether everything is going the way you want it to go. "Do I want my horse to go forward in response to a light leg aid? Then I should focus on these behavioral responses and not use a whip. Of course, I can use a whip to get the horse to go forward, but then he won't offer a sensitive leg-yield or piaffe."

Tools like these bring us back to the realm of trial and error: maybe it will work but maybe it won't. If we use a tool in accordance with the learning theories, we should always think about whether we are just using it as a means to an end and whether we are prepared to accept its possible negative side effects.

With EBEC, we define all of the stimuli ourselves in order to get the correct behavioral response. No matter what training goal we are pursuing, we have the right starting point, provided Level 1 of the EBEC Pyramid is satisfied (the horse is emotionally and physiologically able to respond to our requests). Other tools, such as earplugs that are used because noises make the horse nervous, or blinkers because visual stimuli trigger defensive behavior, or twitches because the horse's fear of syringes leaves him unable to stand still, are symbolic of stimuli that the horse has not understood.

Our goal is to define and only use what the horse *has* understood. EBEC uses relatively little equipment and few tools, which is made possible by guaranteeing Level 1 of the Pyramid. I can't jump to Level 2 when the horse displays defensive behavior or aggression. I never leave the level of learning because there would be a high risk of incorrectly conveying information. In the end, the ones that suffer when we rush are us and the horses we are training.

Expressions of Sensory Well-Being

Horses are able to express sensory well-being. This is one of my favorite ethograms because it shows the harmony that can exist between human and horse. It also allows us to establish instinctively triggered well-being when positive reinforcement is applied in accordance with learning theories.

Horses have been proven to use inter- and intraspecies communication to show how they feel when being stroked, scratched, rubbed, or groomed.

It doesn't matter whether the horse does it himself or whether others help him. The expression of sensory well-being is characterized by the elongation and action of the horse's upper lip. The eyes look to the sides and might also be slightly closed and the ears are usually pricked. As the pleasure increases, the horse may extend his upper lip even farther and it begins to twitch quite rapidly. The horse's nostrils are not flared but they wiggle slightly back and forth in connection with the movement of the upper lip. If the upper lip comes into contact with anything, us or another horse, it will tremble slightly and press firmly into the contact. Then the horse will stretch his head a little farther and might even tilt it to the side. His breathing becomes deeper and he may make a slight grunting sound, while leaning into the scratch. Sometimes, the horse will groom the human at the same time, which is a wonderful moment.

Does reading the above give you the same sense of well-being that I got from writing it? If it put a smile on your face, then you are ready for EBEC! I don't like to see horses rearing, neighing, pawing, and nervous. I know that if I am uncertain, then the horse will be. But if you love this feeling of well-being with horses, then you can aim to get it in training scenarios, too. Level 3 of the EBEC Pyramid (see p. 125) describes how to get there.

"Snapping"—"Licking and Chewing"

"Snapping"—when the horse snaps his mouth open and shut—is an important gesture. A lot of work is done with "licking and chewing" on the natural horsemanship scene. We discussed this in detail back on p. 65. It was and still is defined as a gesture of humility, which also corresponds to the scientific definition. Snapping and licking and chewing occurs when the horse submits, but usually also indicates the beginning of a stress response.

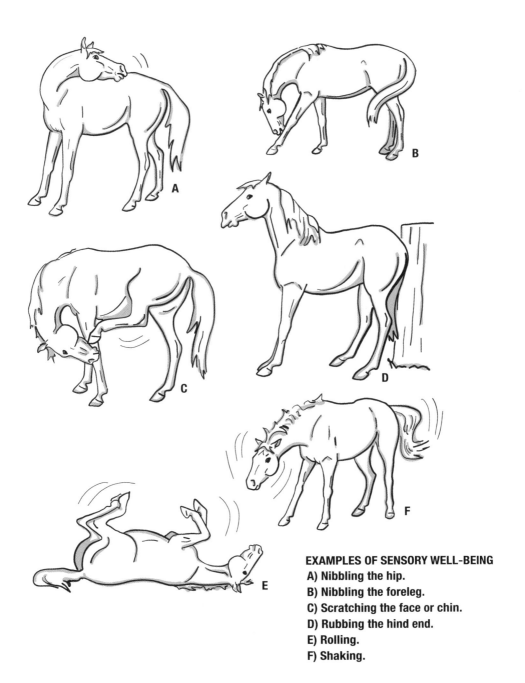

EXAMPLES OF SENSORY WELL-BEING
A) Nibbling the hip.
B) Nibbling the foreleg.
C) Scratching the face or chin.
D) Rubbing the hind end.
E) Rolling.
F) Shaking.

Something triggers concern in the horse, and he consequently shows this gesture. The stress isn't necessarily at an uncontrollable level, but the horse's mouth still becomes dry. The horse only relaxes again when the stimulus that triggered the stress is removed. In the "licking and chewing" that I call "snapping," the horse extends his head and neck. The mouth opens slightly and the corners are pulled back. Vertical movements of the jaw involving an entire range of chewing movements. The lips usually remain slightly parted to protect the soft tissue when the teeth click against each other in the "snap." You can sometimes hear the teeth chattering together. In some cases, only the lower incisors are visible.

Young horses often make this gesture to older horses if they are concerned or anxious about approaching or about their response to something they have done. The behavioral response of licking and chewing is also shown to human beings, cows, or other large mammals. The position of the ears varies, but they tend to be to the side. The eyes are usually looking at the stimulus that is responsible for the cause for concern.

Humility is therefore expressed after previous anxiety or fear. When we see the gesture when training or longeing, it's good to be aware of whatever we have just been doing and what it caused in the horse.

We will see that swinging the lead line, pushing too hard, or using a longe whip causes anxiety. We now need to weigh what to do. We either reflect on it, stop using this stimulus, and allow the horse to relax—for example, by letting him walk—or we finish training and adapt the plan with its positive and negative consequences. If this expression happens while working the horse at liberty in the round pen then we need to calm the situation down, because the gesture might have been in response to us triggering the flight instinct. We should definitely change the direction and bring the horse into the middle. The expression indicates the horse feels driven, both emotionally and physically, and is looking for a way out of his situation.

If we just keep going and ignore the gesture, it will probably be followed by lowering and eventually circling the head. We want to avoid this. We need to have stopped or changed our plan by this point *at the latest*. The first stress messengers should be avoided in the learning process.

Good, competent, two-way communication is only possible if we pay attention. The message is created by how the recipient receives it, and a mind that feels secure is better able to learn than a stressed one.

Gestures with the Tail

In interspecies dialogue with the horse, his tail also gives us plenty of information about what is going on in the equine psyche and body. It should always be included in the overall assessment of physical expression. It is accompanied by gestures with the face, head, and neck, and leg movements.

Tail carriage in mares and stallions during courtship is a language all its own, but that isn't what this book is about. Horses also show numerous movements such as twitching skin or swishing tail movements when they are trying to get rid of things that annoy them, such as insects or itchy skin.

Looked at in isolation, the tail also reveals lots of other information—for example, it can tell us about a mare's cycle if the mare moves it to the side when she is touched. However, that is more relevant for breeding and not important for us here.

A relaxed horse carries his tail low, and it hangs down loosely. If the horse clamps his tail hard against his hindquarters, he is either standing out in a storm by himself or showing extreme humility that must have been preceded by a very stressful situation. This is also seen when the horse tries to take flight to escape an attacker in extreme fear or tries to withdraw from a situation. When the horse speeds up and his stride lengthens, he will raise his tail slightly to allow his legs to move. When the horse is moving

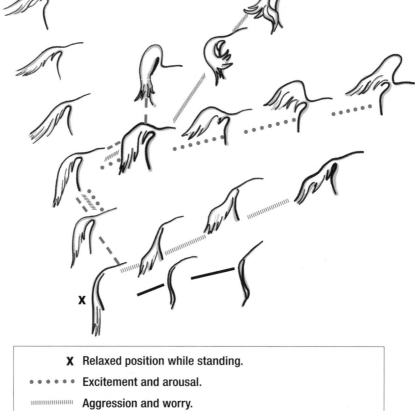

- **X** Relaxed position while standing.
- ••••• Excitement and arousal.
- ⦀⦀⦀⦀⦀ Aggression and worry.
- ▬▬▬ Used in extreme fear, submission, or prolonged pain, as well as when facing downwind during inclement weather.
- ⁄⁄⁄⁄⁄⁄ Used when swishing at insects and prior to kicking, striking, bucking, and balking.
- – – – Used during urination and copulation, and a display typical of a mare in estrus.

The ethogram shows some common behavioral tail expressions. Note that only a few specific behaviors are labeled. For the complete ethogram, see p. 169.

extravagantly, the top of the tail will be almost vertical as the long hair below flows in the wind. Keep in mind that, when his tail carriage is high, the horse is intending to accelerate. When his tail is low or clamped in, the horse intends to slow down.

In aggression, the dock of the tail will point stiffly back, even when it is slightly raised. Powerful sideways swishing of the tail that often goes hand in hand with a strong vertical swing can be observed in horses who are annoyed or irritated. It is usually the preliminary gesture to kicking, bucking, or spooking.

I often find that a powerful tail swish is the first signal when a horse is disturbed while eating in his stall, so I would leave the stall and let him eat in peace. It doesn't matter to me if I have to wait five minutes to do what I want to do. It's an instinctive behavior. The horse is expressing his annoyance with the person disturbing him. In intraspecies communication, another horse would move away and look for somewhere else to eat. We don't see fights over such trivial matters.

When a person ends up getting kicked in this kind of situation, she is often particularly hard on the horse in the next training session, but from the horse's perspective, the two things are entirely unrelated. When this happens, it can destroy a good relationship and the communication with the horse, who will become increasingly unsure of the human interaction from then on.

If we have a legitimate reason for stopping the horse from eating, then we should take a halter with us, walk straight up to the horse, and bring him out of the stall. The tail swishing will stop immediately and so will any behavior related to the food. The horse won't think: "Oh, now she's taken away my feed because I can't be ridden with a full belly and she hasn't left enough time. Well, I'm going to let her have it today. I'm not going to be doing any flying changes. I'll buck her off because she took away my lunch!"

Our human strategic thinking enables us to come up with this interpretation. But the structure of a horse's brain doesn't allow him to plot revenge as we humans might. Most misunderstandings with horses are caused by us acting from a human perspective. We often lose our composure, and what erupts out of us can cause long-term problems with horses.

When a horse swishes his tail, it is worth looking to see whether we could help him by removing what is annoying him. I have seen many people using a whip to swat away horseflies. From a human perspective, this is a nice gesture; from the horse's perspective, it is a painful stimulus that seems like a punishment. Horses are unfortunately unable to reason that this gesture was well intended. This kind gesture can result in a fear of whips or overall uneasiness when the rider sits on the horse. Well-meant can often mean the opposite of good.

HOW THE TRAINING SCALE STILL APPLIES

As we have discussed, we can classify physical signals and significant acoustic signals using the ethograms. Many of the concepts from all three stages of the evolution of horse training (see p. 14) have left too much scope for interpretation. But I do wholeheartedly agree that it is good to stick to the classical Training Scale when we ride.

Guidelines that enable comparability are necessary if we want to be able to hold competitions. The Training Scale was established by the the German National Equestrian Federation (FN). Every English-discipline rider, and some from other disciplines, too, who wants to compete aims to achieve and demonstrate rhythm, suppleness, contact, impulsion, straightness, and collection in the horse. People want "throughness" in their horse—a horse who has been trained to be a pleasant, obedient, and versatile riding horse in both physical and mental respects.

"Training isn't drilling, but systematic suppling. A horse is 'through' when he accepts the rider's aids freely and obediently." (German National Equestrian Federation, 2012)

That is what the judges want to see. The human is now asked to get the horse there "without force." There is enormous scope for interpretation here.

Is the horse "through" if I use my leg to ask the horse to go forward, and the horse accepts the aid and goes forward in response? What happens if he doesn't respond to my leg, doesn't go forward, and I have to use my spurs? When I was young, I loved my spurs with rowels. I was allowed to wear them, which meant I had "made it" as a rider because you had to "earn your spurs."

From my perspective at the time, I had earned them.

Now, some readers might be thinking in disgust: "Oh, she's ridden with *spurs*!" While others might be thinking: "Oh, she must have been a really good rider!" Is either of these views wrong? No! It's just a question of interpretation and the perspective that we take.

I used to describe horse whispering as force-free communication, until scientific studies revealed that gestures such as licking and chewing and lowering the head are signs that the horse has just experienced stress. In my personal definition, a dialogue based on force-free communication shouldn't involve any stress. For me, the phraseology no longer matched the activities. That is just my own, personal "map." The moment I thought this, it changed my approach to horses and to training, even though it didn't for others. But that's okay, too.

The desire to change things grew in me. To me, the Stage 1 of horse training (see p. 13) did not really jive with the definition of "relaxation" as defined by the Training Scale, especially when I looked at many of the training aids used in the horse industry.

For some people, "without force" might be fundamentally incompatible with riding itself, because we are always asking horses to do things that they would naturally never do. Jumping, racing, polo, trail riding, therapy—I've already mentioned that horses have to do so many things for us that they would never do of their own volition. They are flight animals, herd animals, animals who are driven by their instincts and emotions, herbivores, and good at conserving energy. They are prey, not hunters, and are constantly afraid of being eaten. They do their best to conserve energy, and to always be ready to flee from attackers. They avoid wasting energy on unnecessary movement. I would dare to question whether even one single horse would voluntarily turn up to work as a therapy horse or for a competition if we turned him loose and didn't use halters, ropes, or fences to contain him. But noting that this is true doesn't mean you have to be against these activities.

We need the Training Scale because it is meant to be well-thought-out and "pro horse" and not harmful. The FN only leaves so much room for interpretation. And horse whisperers and followers of natural horsemanship might really want to help problem horses who have fallen through the cracks. Many of them try to explain what they mean in words, just as I am doing in this book, and there are countless possible ways to interpret these words because the message is created by how the recipient receives it. Many people will interpret my words as meaning that I am against riding, while others will say that they make sense. This is communication, and as long as both sides are willing, it will result in a dialogue—and that dialogue could be about *only what is best for the horse*.

When I realized all of this, that the concepts and the actions that result from them can be interpreted in so many different ways, I started to think and work scientifically. The concepts that equine and human behavior seek to express are often the problem in communication, not so much the physical actions that we ask of the horse, because the Training Scale, with rhythm, relaxation, contact, impulsion, straightness, and collection essentially makes sense, as do most equestrian disciplines. Doing it right and adopting a horse-centric perspective gives every horse the best possible chance of being able to understand and implement our wishes and requirements. Because if we keep our horses healthy and feed them according to their needs (guaranteeing Level 1 of the EBEC Pyramid); don't overwork them, but pay attention to how they are behaving; don't make them feel stupid, but take leadership confidently and with a neutral mindset; then there is nothing to stop us from achieving our goals with them.

SUMMARY
EBEC Level 2

- Instinctive, innate, and unlearned communication and exchange of information that works equally well with all horses becomes possible.
- We limit people's freedom to interpret equine gestures, enabling the horse to rely on our physical signals.
- The ethograms show clear two-way, interspecies communication with the horse.
- The horse's gestures and signals contain important information for us. They are the basis for applying the learning theories, because they enable us to understand the horse. There is virtually no possibility of misunderstanding.

EBEC Pyramid Level 3

Learning Theories as the Foundation for Successful Training

My literary research into the subject of "reward and punishment" with horses in communication and training began with the 1791 book by Ludwig Hünersdorf *Die natürlichste und leichteste Art Pferde abzurichten* [The Easiest and Most Natural Way to Train Horses—see p. 14] and ended with writings of the present day. I have examined the current texts and articles from equestrian professionals and for trainers, riding instructors, and grooms, as well as the various horsemanship approaches.

PROBLEMATIC TRAINING STIMULI

From all of the stages of the evolution of horse training (see p. 14), I have come across the following stimuli, both for punishment and reward, that we avoid in EBEC. This chapter explains not only learning theories, but why we don't use these stimuli.

Six Common Punitive Stimuli
- Use of the voice (loud or abrupt utterances, clicking the tongue, rebukes).
- Whip aids (crop, longe whip, jumping bat, dressage whip, driving whip, whip for collecting, punishment, cueing, or with horse tied between the pillars).
- Pressure or pain (pulling on the mouth, on the bit, rough rein aids; pressure on the poll, on the ridge of the nose with chains, ropes, or metal, spurs.
- Hitting (smacking in front of the saddle, punishment with the legs, driving forward aids, use of the fists, violence, use of physical strength).
- Pressure in the form of work (energy expenditure, use of various aids).
- Withdrawal (withdrawal of food and/or water, separation from the herd).

Four Common Rewarding Stimuli
- Touch (stroking, patting, grooming, massaging, scratching).
- Use of the voice (praise, soothing words, calming voice, speaking quietly for reassurance).
- Breaks (taking breaks, finishing work, riding at walk outside the ring or on the trail).
- Treats (reward with feed, treats, sugar, carrots).

What is striking about these common punishment and reward stimuli is that they hardly relate to the horse's instinctive, innate behaviors, and on closer examination, they are almost exclusively stimuli that have already been learned through conditioning. They are not *primary stimuli* that correspond to the horse's natural behavior. Furthermore, they only seem (partly) meaningful from the human perspective, and even less comprehensible from the horse's perspective, because they are rarely found in natural equine behavior.

REWARD AND PUNISHMENT: PRIMARY AND SECONDARY REINFORCERS

As common types of reinforcement, *primary reinforcers* are defined as consequences that lead to satisfaction of basic physiological needs (such as water and food).

Secondary reinforcers are defined as consequences for a behavior that are not essential to life (for example, permission to go and play). There are other reinforcers, but they are not relevant for us.

Both types of reinforcers can cause problems in horses. We can't sit down with the horse and discuss putting on his first saddle with him or offer to give him food or water if he cooperates to our satisfaction. If the horse doesn't get it, he can't "give in" and change his opinion about the saddle. He would just get hungry, thirsty, or ill.

Primary reinforcers are also applicable to humans from birth. Unlike horses, human infants won't accept food if they aren't hungry. If an infant cries because she is hungry, and somebody feeds her, she will stop crying because she is full. If the infant is crying for another reason, she will reject any offer of food, and keep crying until her need for a cuddle or a clean diaper, for example, is satisfied.

We get creative with infants and try everything possible to satisfy their needs—pacifier, burping, playing, singing. We take great pains to bring babies relief, and if we find out what they need, then we have been successful—the infants calm down. When a child is later able to communicate, she will be able to point to food or make sounds indicating she is hungry. When an older child is hungry but is in the middle of something, she can finish what she is doing first and have something to eat later.

This is how primary and secondary reinforcers are used from a *human* perspective.

COMMUNICATION AND BEHAVIOR

Before going further, we need to distinguish between *communication* and *behavior*, which is why there are different levels in the EBEC Pyramid that apply to them. This differentiation is a very significant step in Stage 4 of the evolution of horse training. Communication with horses should not be equated with their behavior, with behavioral consequences, or even with behavioral changes.

Communication is understanding through the use of signals and language—that is, verbal and nonverbal communication. Put simply, every

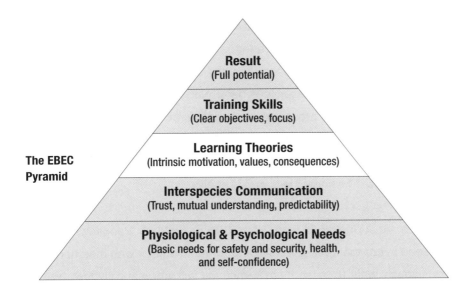

The EBEC Pyramid

communication process could be defined as information being sent from a sender to a recipient. It is therefore an exchange of information, an opportunity for understanding, but not training.

Behavior is an observable action or response, both in people and in horses. Physiological responses, such as sweating, are defined as both a behavior and an experience. Behaviorism or behavioral research, is important for understanding behavior. Behaviorism involves assessing recurring and targeted behavior according to a stimulus-reaction model. If we want to influence and change a horse's behavior, we need to know what behaviors there are and how we can influence them. We distinguish between various behaviors in horses.

- There are movements of the body that we have identified using ethograms and that we can observe and consider in an overall context.
- There are behaviors that can be triggered by external stimuli, including environmental influences. Horses respond to external environmental stimuli according to their natural instinct. This is often innate, unlearned behavior that has a trigger and a response.
- *Operant conditioning* is shaped by the consequences that follow a behavior. These consequences determine whether the behavior is repeated or shown more frequently or less often. If the consequences were comprehensible and pleasant, the behavior will be shown again. If the consequences were unpleasant, the horse will avoid the behavior and it will be shown less or not at all.

Horses are supposed to learn when they are with us. People who want to succeed in competition, are intending to make money with their horses, or consciously want to teach them something will automatically feel this drive. However, this actually applies to anybody who feels they just want to be with a horse, rather than consciously teaching him something. Learning can take place during every encounter.

We distinguish between *intentional* learning (the persistent, continual process to acquire a variety of strategies to improve one's ability to attain and apply knowledge) and *incidental* learning (learning that is unplanned or unintended). This means every process that entails a stable, repeatable change in behavior or a change in how the horse thinks or feels is considered. It doesn't matter whether the information has reached the short-term or the long-term memory. Horses experience things and acquire new skills during the course of their life, whether we want them to or not. It is therefore better to look at learning theories so that we can more consciously perceive and control our life with our horses.

From now on we will be more perceptually aware when we deal with our horses. For example, when you lead your horse out to the field and he takes off when he gets through the gate, he will store this as a learning experience, especially if experiences like this are repeated. If the horse gets a shock from the electric fence in the field, he will store this as a learning experience and avoid touching the electric fence (or any fence) in the future.

If your horse bangs his hip when you lead him out of the stall because you took the necessary bend around the curve from a human and not an equine perspective, the horse has had an unpleasant experience. Your horse will seek to avoid this experience the next time, so he might rush past you out of the stall. If this unwanted behavior doesn't have any consequences, rushing out of the stall will become the norm over time. Learning processes take place even when we are not aware of them and leave us surprised by the new behavior our horse has learned.

If we use a horse for competition, he needs to learn the rider's aids—for example, to go into canter when the rider slides her outside leg back. The horse should also walk into a trailer without stopping. If the horse does stop in front of the trailer, a learning process is underway that we can consciously influence.

A broodmare needs to be led into the stocks and then stand there while she is inseminated. A horse should pick up his feet for the farrier, and so on. All of these are processes that are based on experiences that change the horse's behavior. As human beings, we control them through our behavior, whether we are aware of it or not.

TYPES OF CONDITIONING

There are many different types of fascinating sources of behavioral changes and learning. However, in this book, I only want to focus on two foundations of learning, namely *conditioning*—the process of training or accustoming a person or animal to behave in a certain way or to accept certain circumstances. In the psychology of learning, we differentiate between two basic types of conditioning: *classical conditioning* and *instrumental* or *operant conditioning.* For the sake of simplicity, the latter will only be referred to as operant conditioning.

In *classical conditioning* an association is made between two stimuli during the learning process (see p. 20). A behavior offered entails a consequence. The horse's behavior is therefore not the consequence of a preceding stimulus, but originally appeared at random, including behavior determined by experiences. The consequences that follow the behavior influence future behavior. *Operant conditioning* is described as "reward learning" or "learning through success." The strength of a behavior is modified by reinforcement or punishment—it is either increased by reward or decreased due to punishment.

Classical Conditioning

The process of classical conditioning consists of linking a previously neutral stimulus with an unconditioned stimulus that triggers an unconditioned reaction. When it is paired once or more with a biologically relevant stimulus, a previously neutral stimulus triggers a behavioral response in the absence of a biologically relevant stimulus.

For example, I am starting a young horse and I sit on him for the very first time. Pressure from the rider's legs is a neutral stimulus when it does not trigger any response from the horse. However, if the horse moves off in response to the leg pressure, the leg is no longer a neutral stimulus because it triggers a response.

Let's assume that the leg turns out to be a neutral stimulus for this horse. I will then look for an unconditioned stimulus—a stimulus that triggers a response without conditioning, without learning. An unconditioned stimulus must trigger a natural (innate) reaction (which is why it is called "unconditioned"). My young horse in this example responds to a person walking with him on the ground raising her arm with an open hand. The movement triggers (as interpreted by me) the flight instinct, because in the wild the hand and extended arm are associated with the open claws of a predator. If the open hand does *not* trigger an innate response, such as the flight instinct, in the horse, it has already stopped being an unconditioned stimulus for this horse. Not taking flight has already given it a conditioned meaning. I would then have to look for another way of triggering the desired behavior.

The horse in this example, however, is a sensitive, green horse whose flight instinct is triggered by the open hand. The open hand is therefore an unconditioned stimulus that triggers an unconditioned response—namely the flight instinct.

Classical conditioning is always based on a natural stimulus-response pattern like this (open hand = flight response). The neutral stimulus "leg pressure" becomes paired with the unconditioned stimulus "open hand," the consequence of which is the flight response (forward movement). If the neutral stimulus and the unconditioned stimulus are paired several times, the unconditioned response becomes a conditioned response to "open hand" plus "leg pressure."

In classical conditioning, pairing of the unconditioned stimulus and the neutral stimulus is described as *reinforcement*. The more frequently this reinforcement occurs, the more solid and stable the formation of an association between the two stimuli "leg pressure" and "open hand." If the two have been successfully paired, the previously neutral stimulus "leg pressure" becomes a conditioned stimulus that triggers the desired conditioned response "flight/forward movement" in the horse.

The process is always the same, no matter what we want to teach the horse with the paired stimuli. When training horses, this is necessary for almost all stimuli, because almost all of the stimuli that we use on green horses that have not yet been "confused" by people, are neutral. That includes almost all of the rider's aids, including rein aids.

When the pattern below is followed consistently, without making the horse nervous or putting him on too high alert, he can be rideable and ready for further training within just a few sessions. We need to internalize the following process:

1 Define a neutral stimulus.

2 Define an unconditioned stimulus.

3 Examine an unconditioned response.

4 Pair the neutral stimulus with the unconditioned stimulus so that the unconditioned response follows.

5 Enjoy the conditioned stimulus with the conditioned response.

6 Keep training! Happy horse, happy trainer, happy owner!

Another advantage of this process is that conditioning is no longer exclusively dependent on biologically relevant stimuli because it also works with previously conditioned stimuli, as in higher-order classical conditioning. Behaviors become controllable with an unlimited repertoire of stimuli as soon as they are associated with other stimulus events whose effectiveness is either natural or learned.

Operant Conditioning

Operant conditioning is a paradigm of the behaviorist psychology of learning and involves learning stimuli-response patterns from behaviors that were originally spontaneous. This spontaneous behavior is sustainably changed by a pleasant or unpleasant consequence.

Operant conditioning is also known as "learning through success." The name "operant" is derived from the concept of actively "operating" your environment through actions—that is, influencing your environment through your own behavior. "Influencing the environment" or "being effective in the environment" means that the horse shows an active, physically visible behavior that triggers a response in his environment. The behavior shown therefore has a consequence that in turn influences the future actions of the horse.

We focus on *positive* and *negative reinforcement* in operant conditioning. We want to link responses and their consequences. It is important that the behavior is not the consequence of a previous stimulus but that it originally occurred spontaneously. This is also known as *emitted behavior*—behavior that is not influenced by, or dependent on, any external stimuli.

The consequences of the behavior influence future behavior. That behavior is then the operant behavior. An operant, conditioned behavior, so to speak. The consequence of the behavior must follow directly and regularly. Any kind of dilution, time delay, or distraction can confuse the horse and cause the conditioning to fail. This is not because of the horse, but because of our timing, our stimuli, or the environment.

Operant conditioning therefore intends to change or condition a behavior in the horse. It gives us two extremely successful options for applying consequences that are exceptionally comprehensible to the horse:

1 Positive reinforcement: A horse is more likely to show a behavior as the direct consequence of the behavior shown is pleasant.

2 Negative reinforcement: A horse is more likely to show a behavior because a consequence that the horse finds unpleasant or aversive (not painful or frightening, just "unpleasant") is removed when the desired reaction is shown.

The learning theories also include *positive* and *negative punishment* that make a behavior less likely to be shown. However, these consequences are only successfully applicable to horses in very few exceptional situations, if at all. For the sake of completeness, I will briefly mention positive punishment: when a behavior has a very unpleasant consequence. For example, a horse might get a shock when he touches an electric fence. The consequence is that he avoids touching the electric fence.

Reinforcement is extremely important for EBEC because it enables the horse to offer behavior free from pain and stress, and the horse learns to offer and show the desired behavior without excessive production of stress messengers within his system.

An Example of Positive Reinforcement in EBEC

A horse is learning to be quietly led out to the field or into the indoor arena by himself. The horse becomes nervous when he has to leave behind his friends in the stable. The physical manifestation of this anxiety is in accordance with the "alarm" ethogram (see p. 107) and the anxious behavior begins 30 feet away from his stall. I don't continue to the point where the horse dances around, sweats, or begins to rear. I only go far enough for the horse to tell me with his body language: "Only this far, then I feel unsure and afraid of being away from my herd." I need to offer the horse something that he finds pleasant, in return for him walking 30 feet away from his friends with me. The behavior has a consequence that this horse finds pleasant at that moment.

I determine how far I go—the ethograms help me see how far I can take the horse out of his comfort zone so that he is able to experience the pleasant consequences for the correct behavior. I can read the appropriate distance in his behavior. This takes a little practice. If I go too far and turn back too early or too late, my mistake will become clear the next time we try the exercise. The horse isn't supposed to get the pleasant stimulus when he is upset, but when he is calm and cooperative. This is a question of timing. The positive reinforcer—the pleasant stimulus for the behavior of cooperating—is returning to his friends. If I see the "licking and chewing" gesture (see pp. 65 and 112), I will take the horse back to his friends after a distance of 29 feet on the second repetition of the exercise. But on the third repetition I probably won't turn back until 50 feet. Three to five repetitions are recommended.

If the horse's behavior improves in the next training session, I am on the right track. As soon as the horse associates the pleasant consequence of "being taken back to his friends" with the behavior of "going with the human," the horse will be able to go into the indoor arena or out to the field without stress in just a few training sessions.

If the horse continues to be treated pleasantly when he is there and his expressions of anxiety and fear are acknowledged, then the foundation exists for a partnership between horse and human. In the future, the horse will do everything he can to quickly pick up the exercises the human asks him to do and get them right, so that he receives positive consequences for his behavior.

An Example of Negative Reinforcement with EBEC

The horse is standing quietly next to the mounting block. However, he takes a step forward as soon as the rider puts her foot in the stirrup. The rider clambers onto the horse as he walks. This potentially dangerous situation can be changed using negative reinforcement. In this example, the stimulus that is unpleasant/aversive from the horse's perspective is the rider's toe in the stirrup. It triggers the behavior "moving off."

The aim is for the behavior of "standing still" to increase. This means that the aversive stimulus "foot in the stirrup" may only be removed in the learning process when the horse is standing still. Again, timing is important. Removing the foot from the stirrup too early will not lead to success, because the horse has still not shown any sign of moving forward. However, the unwanted behavior, the horse moving off, is reinforced when the stimulus "foot in stirrup" is removed as soon as the horse has moved off. We need to closely observe when the horse physically expresses that he finds the foot in the stirrup to be unpleasant.

This is easiest when the horse is standing saddled and bridled in accordance with his "neutral" state. The rider stands on the mounting block and looks at the ears, neck, and poll. If in doubt, the horse's attention will be directed to the rear or possibly the side (see pp. 102 and 103) at the moment when the rider slowly moves her foot toward the stirrup. It is not always possible to distinguish between the gestures at first glance. Close observation trains the eye. During the first attempt, gathering nonverbal information is recommended and, during the second attempt, remaining just before the horse's response of "moving off" from the mounting block is recommended. The horse might raise his neck or head or change the position of his ears. He will briefly tilt one ear forward just before he moves because he needs to check whether the area in front of him is safe and that he can move forward.

It is important never to forget that horses do not have a human's ability to think strategically. In our example, the ear tilting forward precisely shows the moment when the rider's foot (because it is the aversive stimulus that the horse finds unpleasant) has to stay exactly where it is. The stimulus is not intensified, because this would cause the horse to move off, it simply remains. The horse will perceive this brief pause. His sensory brain is working at full speed. He notices that the stimulus is not getting any closer and he will go back into the "neutral" position. *That* is the moment when the stimulus "foot in stirrup" that the horse finds unpleasant and aversive, must be removed. If that was a big achievement for the horse, the rider can get down from the mounting block and give him a stroke. (If the horse enjoys being touched then this adds positive reinforcement.) If the horse has already been standing calmly for a few minutes, leading the horse in a small circle and bringing him back to the same position can also be a positive reinforcer.

When the rider repeats the process, it should stay the same. Nothing changes. Within a few (three to five) repetitions where the stimulus is intensified each time, the horse will be reconditioned. He will eventually stop experiencing the foot as an aversive stimulus because the rider mounts carefully and doesn't hurt him.

It is possible that the stimulus "foot in stirrup" had previously been stored as an aversive stimulus in the horse's experience, because the rider, unaware of the learning theories, poked the horse's ribs with her toe when mounting. This can be similar to the driving aid or might even be painful for the horse, depending on the position of the foot. The horse was conditioned to walk forward because the toe was removed from his side when the rider's foot turned forward as she mounted, as soon as the horse moved off. The behavior was therefore conditioned with negative reinforcement. Since horses will do anything to avoid aversive stimuli, this horse might now quickly associate *other* stimuli—for example, moving off as soon as the rider takes up the reins in preparation to mount or raises her knee to put her foot in the stirrup. This depends entirely on the horse's perception at the moment and the "successful" association.

We determine what the horse does. If you don't think about the learning theories, you can unintentionally and unwittingly reinforce the wrong behavior. If one of the aforementioned punitive stimuli is then added as a punishment (see p. 123), the horse becomes stressed and has no way of understanding the process. When you have previously successfully conditioned "moving off" in response to the stimulus "foot in stirrup," reconditioning this behavior with a punishment would be absolutely impossible from the horse's perspective. It can only be done with negative or positive reinforcement.

AVERSIVE AND PLEASANT STIMULI FROM THE HORSE'S PERSPECTIVE

There is another extremely important aspect that makes Stage 4 of the (R)Evolution of Horse Training significant (see p. 125). As already mentioned, we have revised the common use of rewarding and punitive stimuli. We have completely reconsidered them to make those that are used easier for the horse to understand, and above all, to avoid situations of conflict, fear, and stress as much as possible, in favor of a more receptive horse.

We cannot define from a human perspective whether something is pleasant or unpleasant for a horse and can successfully be used as a stimulus. We need to check whether and to what extent the horse is receptive to this stimulus, and also how. Something that is pleasant for one horse, like being stroked or scratched between his eyes, might be unpleasant for another. If he turns his head away and doesn't want to let you touch him, you must not persist. Pulling the horse around by his head or rope and maintaining that this is the chosen form of reward that he must receive is not an option. The horse will clearly not experience and accept this behavior as a reward. Strictly speaking, it actually could be used as an unpleasant/aversive consequence! Good powers of observation are part of successful EBEC training! A horse only perceives being stroked between his eyes, or on his neck, head, or withers as pleasant if he behaves in a way similar to the "well-being" ethogram (see p. 113).

It is also important to ensure that the training environment is quiet and free from distractions. If the horse's attention is elsewhere, his behavior might not change or a behavior other than the desired one might accidentally be reinforced.

Seeking/Avoiding a State of Pleasure

Edward Lee Thorndike, who is known for his work on learning theory that led to the development of operant conditioning, did not understand the concepts of "pleasant/unpleasant" and "appetitive/aversive" as subjectively experienced conditions—as such they would have no place in a behaviorist theory. Instead, these concepts can be categorized according to whether the animal seeks or avoids them. Thorndike defined it as follows:

"responses that produce a satisfying effect" are more likely to occur again in that situation and "responses that produce a discomforting effect" are less likely. A "pleasant" condition is a condition that the animal does not avoid but often seeks and maintains. "Unpleasant" describes a state that the animal normally avoids or leaves.

Any behavior that is shown in a certain context might be performed more frequently under the same circumstances in future. As an observer you can then assume that the consequence was "pleasant" from the horse's perspective. If the behavior is rarely performed, the consequence was probably "unpleasant" from the horse's perspective. The concepts of "pleasant" or "unpleasant" in these circumstances are not defined by people but by each individual horse.

If my horse really likes carrots and I give him a carrot for "sitting down" when we are working at liberty, and then, within just two or three repetitions, he always sits down because he knows he gets a carrot when he does, then this horse experienced this stimulus to be pleasant in this situation. Another horse might accept the food, but won't sit down more frequently, and might even start to nip instead. In this second case, the horse has not stored this stimulus as pleasant for this training goal.

I gather demonstrable evidence that either confirms my approach or not. However, I have to make sure that the carrot is not the stimulus that triggers the sit. The stimulus can be something different such as the verbal command "sit" or a tug on the halter. The carrot is the reward, the positive reinforcement for a result that comes closer to the goal.

For a horse who, for example, is supposed to learn to be in the arena by himself, the best reward or the biggest incentive, at the moment he has managed to calmly walk a small circle in the arena, is to be allowed back to his friends in the barn. A carrot or other treat probably isn't an incentive at that moment. I can clearly tell this by whether the horse shows the same result on the second repetition of the exercise or even improves and can calmly walk in two small circles. It is important to observe the horse's gestures and refer to the ethograms here. I can bring the horse slightly out of his comfort zone but not so far that he is afraid.

DISTINGUISHING BETWEEN INNATE AND LEARNED BEHAVIOR

When we train horses, we need to recognize what they can already do and which external stimuli make this behavior visible for us. An innate behavior is triggered by the presence of a corresponding stimulus, whether or not we want it to be. The horse has no cognitive influence over whether or not he offers this behavior. It is triggered naturally. It is therefore very important to distinguish between *innate* and *learned* behaviors. We can recondition the latter if we want to; we have to accept the former because they cannot be changed. However, we can make excellent use of them in classical conditioning as unconditioned stimuli to develop a conditioned response, and also as reinforcers in accordance with operant conditioning.

Innate behavior is inherited behavior. Behavior refers to all movements of a healthy horse as defined in Level 1 of the EBEC Pyramid (see p. 34). Innate behaviors are those that cannot be changed though learning processes. You can tell that a behavior is innate because an innate response action is triggered by an external *sign* or *key stimulus*—a "releaser" or the determining feature of a stimulus that produces a response.

There are external stimuli that trigger a behavior in an individual horse. For example, a flapping plastic bag can cause one horse to spook but not the horse next to him. The plastic bag is therefore an external stimulus that doesn't influence the behavior of the other horse, so it is a highly individual, triggering stimulus. However, there are also external stimuli that trigger action in the second horse as well—the alarm sound "blow" (see p. 76) is an example of a stimulus that all horses respond to with a state of alert.

Investigations into the social behavior of horses, which are herd animals, show the extent to which we need to distinguish between instinctive and learned behavior in order to be able to use them for our benefit in training. If a herd is standing in a storm, and the wind is swirling leaves, dust, and maybe even plastic bags around, the horses will rarely group together. They will usually be scattered far apart, eating or standing with their hindquarters into the wind. However, the activation of a single sensory stimulus, such as the sound of cracking sticks as "something" creeps closer to the field, can cause all of the horses to move closer together for protection against the potential danger.

For the sake of completeness, I should mention that it is also correct to differentiate behavioral responses into *instinctive* and *reflective behavior* (the latter being behavioral decisions that are based on knowledge); however, in this book, we are only focusing on conditioned/learned and instinctive/unlearned behavior, to enable us to train successfully.

The application of EBEC requires us to closely observe and look at the horse we are working with and whose behavior we want to influence, without interpretation. We will be richly rewarded when we are able to develop an efficient stimulus concept from physically visible behavior that is highly individual to each particular horse—that is, we are able to derive appropriate positive and negative reinforcers from the information given. The horse in my earlier example, who is supposed to learn to calmly leave his herd and walk alone in the arena through positive reinforcement, experiences a positive reinforcer after having successfully completed the task: the company of his friends (what he longed for most at this moment). The reinforcers we use must not be too abstract. "Abstract" in this case would be something like: "You walked with me calmly in the arena. As a reward we're not going to work in the indoor arena today, which you don't like, we're going to go outside for a ride."

The positive reinforcer must immediately follow the behavior, and the horse must be able to associate it with the behavior. So even if I did give him a treat at that moment (some tasty apples, for example), he wouldn't be able to associate this food reward with leaving his friends in the barn. After all, we have taken him slightly out of his comfort zone. He will prick his ears, listening for the herd. He will be tense. Horses aren't usually accepting of food when they are tense. If a whinny from another horse in the stable activates my horse's auditory sensory brain, it will be virtually impossible for him to associate the treat I want to give him with positive reinforcement. The temporal proximity (how close in time thing occur) between the behavior and the positive consequence is particularly important in horses, because they are unable to think through situations strategically. They are therefore incapable of thinking: "That was a nice ride. So tomorrow I'm going to be a bit calmer when I come out of my stall and leave my friends to go into the indoor arena!"

We are focusing on a behavior of the horse that is triggered by an external stimulus. The horse's *sensory perception* (see p. 56) triggers behavior. We have learned this is different from human perception. This is one reason why training methods where people try to imitate the behavior shown by horses when they interact among each other are not always successful. Adaptation is virtually impossible, as well as unnecessary.

Instinctive Reinforcement—The Heart of EBEC Training

Just knowing how horses can express themselves through body language, and knowing what their sensory perception is capable of, doesn't tell us whether a specific stimulus will really trigger a certain response in the horse. We might be able to trigger a flight response with an open hand, but the horse might stay at the same pace in response to our open hand, because this stimulus is either conditioned or unconditioned in this horse.

It is important to know what horses can and can't perceive with their senses—for example, which changes they can see in the environment that we humans can't see (review the diagram of the horse's field of vision, p. 101). We can only see in front of us, but horses' all-round vision enables them to perceive visual stimuli behind us, and as we've discussed, this can trigger behavior in the horse. However, the fact is also that not every horse responds to every change with a movement, even though his sensory organs are absolutely able to see, for example, the women with the umbrella behind him. Not every perceivable stimulus simultaneously triggers a response.

In his investigations into instinct theory, Professor N. Tinbergen put it very clearly:

"Effective stimuli are established by trying all possible combinations of stimuli. Instinctive action normally only responds to very few stimuli. Everything else that the environment offers is almost or entirely unimportant, although the animal has all of the senses to perceive much of it."

This describes very aptly why rigid standard rewards and punishments are not reliably effective for training the behavior of horses. If I make a sound by clicking my tongue that is similar to an approaching predator treading on a twig, and it does not trigger instinctive flight behavior, this stimulus is not effective as a forward aid at that moment. A tactile or visual stimulus might have a better triggering effect. We therefore need to distinguish between:

- What the horse can perceive.
- What he actually perceives at that moment.

- What he actually responds to in the specific training situation.

I cannot emphasize often enough how important it is to know the difference between innate and learned behavior. Trained, conditioned, learned behavior can be retrained at any time. Innate, instinctive behavior, however, can always be triggered and does not have to take any "detours." In a nutshell: Innate behavior stays; learned behavior can change.

How can we tell whether the behavior our horse is offering is innate or learned? Lazy horse one day; excessively motivated horse the next? If we regularly observe our horse going much faster out in the field with his friends than he does when we ride him in the indoor arena, that means that our horse is able to go faster. This corresponds to his innate ability and is breed specific. It does not mean that the horse is lazy but that the stimuli being applied under saddle are not triggering the desired behavior but a different one. However, it also does not mean that the horse would not be able to show the desired behavior in response to a corresponding triggering stimulus—it just needs to be conditioned appropriately.

Horses only respond to a few key stimuli. A horse might not respond to a key stimulus in the indoor arena, but he might respond very well in a different environment like the outdoor arena. When I was developing EBEC I initially tried to find out which key stimuli triggered certain behaviors. Which key stimuli are most important for us in communicating with and training horses, and which can we use in order to be able to classically condition a horse?

We know from brain research studies into horses that strategic thinking has little or no influence over behavioral responses that are controlled by instinctive behavior. That means that a horse is not able to reflect on whether or not a spook is really appropriate at a particular moment. He is not able to suppress the behavior intellectually. The spook is triggered instinctively and directly by an external stimulus. Once an instinctive action has been triggered, we have to deal with it, and EBEC gives us a way to do this. We can even use it purposefully for our benefit. The instinctive reaction is directly connected to parts of the horse's brain and the resulting physical movements made by the horse, without any loss of time or any distractions, both when the triggering stimulus is triggered and when it is removed.

I explain this possibility using the example of positive reinforcement on page 132: In order to teach the horse to gradually move farther away from the stable and to allow me to lead him away from his herd, I must be in tune to the moment the instinctive action of "social herd behavior" is triggered in the horse. This is when the horse can still tolerate the pressure on an emotional level. I remain with the horse, watching to see whether his physically observable behavior changes in accordance with the ethogram (see p. 91). Can the horse "get to like" the situation of being away from his herd, and can there be a certain type of "relaxation" and acceptance of the situation experienced, in the emotional brain? The answer is yes, at this precise moment, which can be created and controlled in every healthy horse (provided the horse is in the situation because of the trainer and not the horse himself), when the horse experiences a positive reinforcer (the desired relief—return to his herd). This is how to create an *instinctive reinforcer*. The horse gets what he longs for most: He is allowed to go back to his stablemates. I am therefore working with a stimulus that is innate in the horse and therefore fully comprehensible to his entire being, and creating a stimulus-response behavior that is effective within fractions of a second and that has a conditioned behavior as a consequence. Namely, the horse will allow me to "distance him" from the herd.

I can now gradually intensify the stimulus, and the next time he leaves the barn, the horse's emotional brain will say that "being calm is useful," because it worked well for him the last time. This isn't cognitively thinking through the situation; it is the direct result that the horse experiences through his behavior that was controlled by our response. It is important to identify the moment when the horse shows the still unconditioned behavior, "I can't leave my herd." He must not reach uncontrollable fear, otherwise instincts will entirely take over the horse's body and he will not be able to "consciously" process the situation.

The end result is that the horse learns in as stress-free a way as possible to quickly find his way around a situation that is contrary to his nature. The horse's instinctive behavior becomes a toolbox of possibilities that we and the horse will always have with us and that costs us nothing. It is a creative toolbox that can be read from the horse's behavior. We can derive instinctive reinforcers from flight behavior from social herd behavior that in itself is so varied and multi-faceted that it gives us endless possible ways to directly reach the horse. We can also learn from and use "into-pressure syndrome"—the innate, unlearned behavior in horses that switches on the first time there is pressure from the rider's leg in the belly area. It says to the horse: "Don't do anything, keep moving into the pressure, and wait and see what happens. If it's a predator, its claws might loosen and you can run away." Remember that just about everything that we teach horses in today's civilization does not correspond to their natural behavior. We have "designed" life with horses from *our* perspective. If we use the horse's instinctive behavior as an efficient reinforcer in order to bring about a long-lasting change in behavior, we will be working far more efficiently from the horse's perspective.

Punitive Stimuli Example

Horses learn to learn and have been proven to continuously improve with every learning situation in which they either successfully receive positive reinforcers or successfully avoid negative reinforcers. Once they have internalized this simple concept, they will quickly offer various behavioral responses, and it is up to us how easy to fulfill we make them. As mentioned at the beginning of this chapter, punitive stimuli aren't likely to bring us success.

In the wild, where horses are free to choose how to eat without our involvement, they eat for 12 to 18 hours, based on a 24-hour daily cycle, chewing 40 to 80 times a minute. A large horse needs 40 to 50 minutes and 2,500 chews to eat just over 2 pounds of hay. He can eat around 2 pounds of oats in 10 minutes with 800 chews. Withdrawal of food as punishment for behavior only causes further behavioral problems, such as cribbing, stall-walking, or weaving. Satisfaction or the promise of satisfaction of a basic need will rarely result in the horse learning to accept his first saddle, bucking less, or responding better to the leg aids during lateral work. Neither will it help him find a good distance in a round of show jumps. If he is afraid of the jump or the rider's aids or the monstrous tractor that has been parked in the arena to make the approach more difficult for a more exciting competition, he will refuse the jump and pick up faults. This happens entirely independently of whether or not his basic needs (Level 1 of the EBEC Pyramid) are fulfilled. In addition, as we have discussed, horses can only save information, learn, and change their behavior when their basic needs *have* been fulfilled.

The horse in question is likely to offer a behavior in response to stimuli and be able to do the exercise successfully when the instinctive behavior triggered is reduced or the opposite behavior is offered, like the hungry infant who stops crying when somebody feeds her. For example, the horse will stop his defensive behavior in response to leg pressure if the leg pressure is given at the right moment for the correct behavioral response and then removed (negative reinforcement). When done right, the desired behavior will increase, not the unwanted one.

We have to show the horse understanding, because it is we who put the horse in difficult situations. We either create defensive or anxious behavior ourselves or expose the horse to environmental stimuli that cause it.

Handling and training from the horse's point of view adds that extra something with horses and solves a lot of problems. Determining reinforcers from instinctive behaviors is therefore an extremely efficient way of applying the learning theories we've discussed. That is why I have dubbed these stimuli "instinctive reinforcers."

Conditioned, Positive Reinforcer Example

Using conditioned, positive reinforcers in horses can bring difficulties. For example, I have found that success is initially seen more quickly in horses who are given a carrot for every correct behavior (they are continuously positively reinforced with food), but the stability of the long-term effect is low when considering the overall outcome. If the positive reinforcer fails to materialize within the normal time, the horse quickly forgets the behavior—or it was never established in the first place.

When working at liberty, the horse might put his head down between his front legs when asked because there is a carrot there. If the carrot isn't there, he might stop offering the desired behavior altogether, or he might threaten aggressive behavior—even biting or kicking. In my opinion, one of the reasons many trainers always carry a whip is to protect themselves against any defensive behavior by the horse.

In this case of the horse at liberty, the behavior has been triggered by the positive reinforcer (the carrot). In my opinion, giving food to flight animals is an ineffective model. People and dogs have been proven to learn more slowly but more sustainably with *intermittent reinforcement*, where the desired behavior is only reinforced with a treat now and then. My experiments do not confirm this is the case in horses, because horses are extremely focused on the carrot and try to find out how they can get to it rather than focusing on the behavior. This is a brainteaser that their brain capacity does not allow them to solve, no matter how hard they try. If the horse only gets a carrot occasionally—for example, every third time he lowers his head—the agonistic (aggressive) behavior is observed even earlier and is sometimes even more intense. (However, to be able to conclusively evaluate this scientifically, other investigations need to be done.) In addition, and as already mentioned, the fact remains that, as soon as a moment of anxiety is triggered—at a competition venue, in the trailer, in the stocks, with the farrier, in an unfamiliar or changing environment—the horse will not perform for food, because after a certain stress level has been reached, the horse will not accept any food.

My numerous experiments have shown me that it is important to provoke intrinsic motivation in the horse who is independent of the reinforcer.

The horse who is worried about leaving his herd or the familiar surroundings of his stall will go with me as far as he can tolerate it emotionally. In so doing he learns, on the one hand, that I acknowledge his behavioral response. He is aware of this because I respond to the change in his behavior—that is, his behavior has a direct influence on me. Because I do not keep intensifying the pressure to the point of being defensive, fighting, or causing stress or aggression, he continues to deal with the situation "positively." We know he isn't capable of thinking this through strategically, but he can experience it physically through repetitions. By keeping the stress level as low as possible, but not quite at zero, the horse experiences, that any concern he felt was not justified. This is because he feels that he got back to his friends despite his concern and that nothing bad happened to him.

If the horse is not punished for taking this risk, but constantly influenced in a sympathetic way, this will increase his willingness to take risks in other situations. "What did I do and how did it work out for me?" now triggers information in his limbic system, in his emotional memory.

By using this method, rather than a conditioned, positive reinforcer, I prevent offering or trying out various behaviors that correspond to the horse's nature, as occurs in the example of the liberty horse who wants to somehow get to carrot. For example, the horse might snap at me if lowering his head between his knees does not result in a carrot. In addition, he will probably not offer the desired behavior again because it wasn't successful the third time. He will snap at me because he didn't get the carrot. The horse offers behaviors that he comes up with himself, things that correspond to his catalogue and "map." These might now include, rummaging around in my grooming box or bag, looking for carrots, or being more aggressive and pushy with his ears back when we try to do the trick. If I give in and give him a carrot at this point, I am training an aggressive horse with negative reinforcement, and the pushy behavior with the ears back will only increase.

Using instinctive reinforcers, on the other hand, trains and strengthens a wonderful ability in the horse. A conditioned action becomes a self-motivated action, because the horse himself finds solutions for improving his situation in life from his very own skills that are available to him. It is not external circumstances or consequences that lead to success, but his changing inner attitude. He learns to influence the scenario with himself. From the horse's perspective, it would go something like this: "My environment perceives me, responds to my behavioral responses, and as a result, I experience that the things I was afraid of and that I didn't understand, are not bad." Put simply, we enable the horse to take his life "into his own hands."

Sensory Memory
Sensory memory refers to the traces of memory that are formed in the sensory area of the central nervous system, every time a sensory receptor is activated. When pressure with the legs or a tap with the whip are given, the receptors activate the tactile sensory memory; in the case of a vocal aid, the auditory sensory memory is activated.

Everything that happens remains in the horse's brain in small traces

Learning success

Joy

Self-confidence

Curiosity

Event or object

of memory. We often simultaneously activate many sensory areas, either consciously or unconsciously. Only around 2 percent of sensory input is estimated to find its way into the long-term memory.

Let's assume that a horse doesn't pick up the canter in response to the leg aid and so he is then punished with the sensory stimulus of a smack with the whip. But at the next attempt, the horse still won't canter. He still doesn't show the desired behavior during the subsequent repetitions. The horse doesn't understand the concept or the stimulus. This is proven by the fact that the behavior doesn't change. But why doesn't the horse pick up the canter?

We know that the *short-term* or *primary memory* (memory that retains only a small amount of information for a short period of time) needs continuous activity that involves the nerves. When something interrupts the process (for example a smack with the whip or an unexpected distraction), the path to the

Failure

Pessimism about learning

Low self-confidence

Decreased motivation

Continued failure

short-term memory is lost. The short-term memory appears to have a temporal sequence—what goes in first is replaced by what comes in next if it is not repeated often enough (such as when a behavior is learned).

In this example where the horse is supposed to respond to the stimulus "smack with the whip" with the behavioral response "canter," the path that we are trying to lay is deleted with every smack, and we have to start again from the beginning. The horse cannot offer the solution because, in this learning process, the receptors are working against it.

The horse might now successfully go forward in response to the tactile stimulus "smack with the whip," but he will not become any more sensitive to the leg. Many riders then reach for their spurs. This creates a vicious circle. It is rare for riders to abandon their whip and spurs after a few successful repetitions.

Horses often develop a fear of the whip and run away. Then, a strong hand on the reins sometimes gives us a feeling of control. In the meantime, though, everything in the horse's sensory memory is out of control, so to speak. It can "work" to use a smack with a whip to get the horse to canter, but on the whole this concept far too often *doesn't* work. The risk of producing a horse who is out of control under saddle, be that too fast or too slow, is simply too great. Positive repetitions with an appropriate stimulus are very important for enabling storage in the long-term memory.

When a horse who is supposed to understand the basic leg aids remains still in response to pressure from the rider's legs or becomes slower and slower the more we intensify the pressure, instead of going forward or going forward faster, he is not unusual. The horse has two problems:

He can't look at another sport horse and copy their behavior, because we have to assume from the latest scientific findings that horses do not have *mirror neurons* (neurons that fire both when an animal acts and when the animal observes the same action performed by another) that work like ours do. They can observe the behavior of other horses but rarely do they imitate it.

And they don't understand the "game" that we have thought up anyway, because if they did, they would find it easier to do.

The concept of moving forward, moving forward faster, or moving forward in a different way in response to a leg pressed against their belly area doesn't exist in the horse's world. Watching other horses being ridden can't enable them to understand riding any better. The horse has no idea what to do with the leg pressure, has never correctly understood a gentle leg aid as it should be, or his instinct is such that it forbids him from yielding to the pressure from the rider's leg. As mentioned, he naturally initially goes *into* pressure from the leg, instead of away from it. This is easy to observe especially when starting young horses. When Level 1 of the EBEC Pyramid is guaranteed and the horse is completely free of fear, this "into-pressure syndrome," as I call it, is clearly observable.

If the horse now experiences a little or a big smack or several or stronger leg aids, the attention is taken away from the actual goal (forward movement). In accordance with negative reinforcement, the leg must remain on the horse's body until he moves so that the horse understands that this is supposed to be a forward aid, contrary to his instinct. A flapping leg or additional aids are not forward aids. This is too complicated for the horse's brain to understand. When we do this, we are back at trial and error again, which very well might not get us to our actual overall goal, namely that the horse understands that he is supposed to go forward in response to a gentle leg aid.

Going forward to leg pressure is what we humans have come up with, so horses need to learn it. It's what the whole of equestrian sport is based on. A lasting change in behavior therefore needs to be established on the basis of repetitions. If the horse feels pressure from the rider's leg in his belly area, it means that he should go forward. He should give this response calmly, to light pressure, because it is likely he will later have to do exercises in response to firmer pressure from the leg. When the horse responds with the behavior of "going forward," we remove the stimulus. We ensure the transfer from short-term to long-term memory through repetitions. This needs to be done clearly and faultlessly and in accordance with Level 4 of the EBEC Pyramid (see p. 157). The communication between human and horse (when which stimulus is applied or removed), must not be weakened and must, above all, not involve any tension.

If we punish the horse with the whip because he is slow or sluggish, or as the vernacular puts it so beautifully, "lazy," and still doesn't move in response to increasing pressure from the leg, we are using a punitive stimulus that is extremely difficult for the horse to decode. We also risk exposing the sensitive structure of the skin to pain. Furthermore, it is virtually impossible for the trainer to precisely repeat this response, because repetitions need to be an exact copy of the previous time for the horse's brain to have a chance of making the connection. If you punish the horse for standing still and a slow response to leg pressure, it is possible that, after he has been ridden for a while, the horse won't understand that he is supposed to stand still for the rider to dismount—even though he can still feel the rider's leg. This is incredibly difficult for horses to understand. A very nervous horse can therefore become conditioned to stand still for neither mounting nor dismounting and is always ridden in a state of "alert," which makes rhythm and suppleness impossible. Even if the horse is very calm and Level 1 of the Pyramid has been guaranteed, a punitive stimulus can trigger all kinds of responses that people are unable to control. The risk of an unwanted behavior is very high.

An exercise only needs to be repeated two to three times to see whether the learning theories have been applied correctly and are working from the horse's point of view. If the desired behavior occurs repeatedly in response to the chosen stimulus, the aim has already been achieved or the way to achieving the aim has been determined. If nothing changes or the behavior deteriorates, the training plan should be changed.

The results of my tests on thousands of horses at Paul Schockemöhle's Lewitz Stud have been clear: When a training exercise doesn't work after a few repetitions, the desired learning success won't be achieved, even after *hundreds* of repetitions.

Timing and Scope for Interpretation

Timing is crucial for adding and removing instinctive reinforcers. If your horse enjoys being stroked, you can use stroking as a positive reinforcer, for example, when leading the horse and he halts beside us when we come to a stop. We can make excellent use of stroking in this scenario.

Scope for interpretation plays the leading role here—it's the same problem that I discovered in my reading and when studying natural horsemanship. We therefore need to limit the scope for interpretation when training a horse. When is something a "touch," a "tap," or a "smack"? What strength and frequency of smack achieves the result and is justified? Is a smack the same for a pony as for a large horse or draft breed? Which behavior is reprimanded at which intensity and at which moment? Are foals punished less severely because they are younger and smaller? How hard am I allowed to push back a foal when he keeps running into me?

In my science-based method that comes from the horse's perspective, I have minimized scope for interpretation as far as is possible, and using the ethograms I shared in chapter 4 (see p. 88), clearly defined the behavioral responses and consequential behaviors in response to a stimulus and classified them according to body language. For example, we can precisely define how long a horse's leg needs to be lifted in preparation for the farrier. If the horse reaches Level 5 on a pressure scale of 1 to 10, and shows this through his behavior, we keep holding onto the hoof. The horse will fidget. If we stay at Level 5 and don't intensify the pressure, the horse will become calmer. In that moment of calm, we can put the hoof back down again. We are rewarding the horse with positive reinforcement. If, however, we intensify the pressure to 6 at the moment when the horse is fidgeting, he will continue to fidget. The intensity of his defensive response will increase. The horse might fall over or rear.

If the horse finds standing still difficult, we can lead him in a circle and bring him back to the starting point to try again. The horse will probably stay at Level 3 to 5 of the pressure scale and check whether we will set down his leg when he remains calm (repeats the behavior). After three to five repetitions, a pathway has been established in the sensory memory.

The horse will be able to remember this in the next training session, enabling us to make massive improvements. Interpretations of what to do and how are kept to a minimum.

In the past, in all stages of the (R)Evolution of Horse Training (see p. 14), we have made it far too easy for ourselves when it comes to the theories of rewards and punishments in horses. They make sense from a human perspective, from the horse's perspective, not so much.

SUMMARY
EBEC Level 3

- In order to reduce or increase a behavior, we teach the horse consequences for his behavior that are meaningfully motivating or demotivating, from his perspective.
- Not every perceivable stimulus simultaneously triggers a response.
- The consequence of the behavior must follow directly and regularly.
- Repetitions are very important for enabling storage in the short-term and long-term memory.
- We recognize the stimulus that is responsible for the change in the horse's behavior and ensure that the horse perceives the stimulus to be "harmless" and "acceptable" as quickly as possible.
- It is extremely important to base rewarding and punitive stimuli on the horse's natural and observable behavior.
- We notice all of the horse's body language signals as soon as he changes from "neutral" to "concerned" or "annoyed" or another state in response to a stimulus.

EBEC Pyramid Level 4

Clear Objectives and Focus

We achieve the most training success when we determine the stimuli we use in classical and operant conditioning from the horse's innate behavior. The stimuli are then not stimuli that have already been classically conditioned and that can only be used in certain situations, but instinctive stimuli derived from the horse himself that are always successful.

It is extremely important to choose rewarding and punitive stimuli based on the horse's natural and observable behavior and not from stimuli that been trained that might not work if the situation changes. It is also very important that the horse always remains below the threshold of agonistic, alarmed, or aggressive behavior, in accordance with the ethograms we studied in chapter 4 (see p. 88). This includes expressions, such as a threatening face, kicking with a hindleg, or the swish of the tail. Horses often swish their tails in a defensive manner when their riders poke them with spurs or use their whips.

Horses don't have the same ability to multitask as we humans. If a horse has to concentrate on defending himself against a stimulus, his attention will be divided, and he will develop defense mechanisms and therefore never be able to perform as well as a completely forward-thinking, attentive horse when he approaches a jump or performs a dressage movment. For horses to be able to reach their full potential, all levels of the EBEC Pyramid must be taken into consideration.

The horse slipping into defensive or aggressive behavior is also counterproductive for regular training at home and for preparing for a new training exercise. It can result in the horse constantly fighting to "ensure his survival," which prevents him from learning a stimulus that is supposed to have a certain behavior as a consequence. In training, this is what tells us when we have gone too far or when a stimulus has been presented with too much intensity.

We need stimuli and the horse's appropriate behavioral responses to them to achieve our goals, whether we just want to ride out on the trail safely and calmly or for general handling in the barn, or in order to work our way through the Training Scale. It is therefore extremely important that we establish the stimulus-response behavior that we discussed in the last chapter very clearly and do not leave it. Hitting, shouting, and causing pain

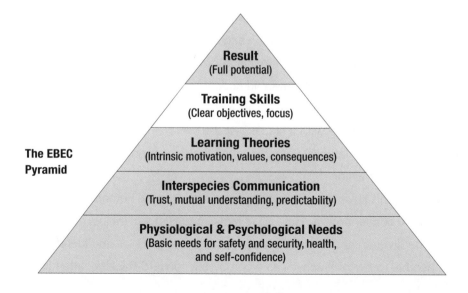

interrupt the learning process, and in the worst case, can even replace what we have already taught the horse in his memory.

Transforming learning into behavior requires motivation. We can create this motivation in every horse by seeing things from his point of view. We also need to set clear objectives and identify our focus on the proper positive and negative reinforcement to achieve those objectives. The stimuli that the horse himself finds pleasant or unpleasant must be derived from his instinctive behavior. The less we dilute the stimuli and our responses to the horse's behavioral responses, the easier it is for the horse. Our horse will thank us for staying true to the EBEC Pyramid.

SUMMARY

EBEC Level 4

- Transforming learning into behavior requires clarity.
- We need to have a clear objective for which behavior the horse is supposed to offer in response to which stimulus.
- For the horse to be able to fulfill our wishes, we need to be focused on what we want and on what the horse does.
- We must be self-reflective so that our actions are based on the facts. Horses aren't capable of following our train of thought; they can only be guided by our actions.
- Horses follow the pattern of: "What did I do and how did it work out for me?"
- If a few repetitions don't bring us closer to our goal, we need to rethink our training.

EBEC Pyramid Level 5

Fulfilling Potential Through Self-Reflection

Many years have passed since I first got the feeling that horses have a communication system that might be foreign to us but that we need to learn to understand. Even during my childhood with the neighborhood Icelandics, I noticed that taking one of them out for a ride without the others went against their nature. I had a feeling it wasn't just about me or being ridden when my pony Dominik reared and took off. It was about going against his nature by tying him in a stall three days a week so he was bursting with energy and wanted to run. I knew deep down in my heart in both cases that the horses' behavior wasn't anything to do with me and that something in the entire system around them wasn't right.

I always knew that horses don't hate us. Whether they can love us is still a question.

During all the years when I was doing natural horsemanship demonstrations around the world and working with problem horses, I was sure that there was still something that we humans hadn't understood. But I didn't know what it was. I did my best to make problem horses into functioning horses, mainly so that they didn't have to die. For most horses who become aggressive, unsafe, or hurt somebody, their last journey is a "one-way ticket." Professionals can't keep horses once they know that their problem behavior will prevent them from finding a buyer. For a whole variety of reasons, pleasure riders aren't able to cope with the kinds of problems that can arise. I suffered for many decades because I didn't have a real solution, and as my scientific research revealed that natural horsemanship and horse whispering didn't have all the answers either, I often despaired. But I always had this confidence, this certainty that I or somebody else would find out more and life for horses all over the world would get better.

Right at the start, when we first get involved in horses as a career or as a hobby, we should make very sure that we always keep this empathy and love for horses in our heart. We should always carry this within us. It's harder to shout at or hit someone you love. No matter whether you are an equestrian professional, competitive rider, breeder, or pleasure horse owner, trust the feeling that you have in your heart. Stress has been scientifically proven to have a negative effect on horses' performance. EBEC helps you to minimize your stress, and your horse's.

People are still looking for solutions, and solutions are what I hope to offer with this book and the EBEC Method. These ideas will definitely develop further, and this definitely won't be the last book. If you get stuck at any point, contact AKA (andreakutschacademy.com). We have now become a wonderful community of people with students, Certified EBEC Trainers, and EBEC Master Trainers who can help you. We offer online coaching; modern media helps to teach theoretical content in webinars, and we can resolve just about any conflict situation using video analyses (see p. 169). And if you want to turn your hobby into your job, we are willing to share with you everything we know. You will become more successful with horses, and faster, and be able to create an impressive, instinctive understanding on the ground and in the saddle that will bring tears to your eyes more than once, as it has done for me.

The most important thing for being able to apply EBEC in the future is self-reflection, because the solution lies within us.

SELF-REFLECTION

Self-reflection is the ability of people to reflect on their own situations. Reflections on external or internal observations can be seen as opportunities to recognize problems and identify starting points for making changes. *"Self-reflection requires the capacity for nuanced introspection and a certain distance from yourself"* (Stangl, 2019).

It sounds ever-so straightforward. But it isn't, because self-reflection is a question of practice. Our subconscious or unconscious frequently gets out of hand. When that happens, we don't control our actions with our cognitive brain. Instead, our actions are determined by our previous

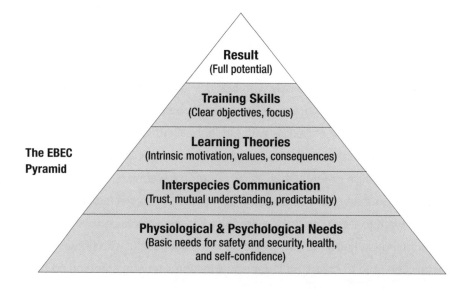

The EBEC Pyramid

experiences that are stored in our system, in our personal "map." As I mentioned at the beginning of this book, scientists say that we could think of consciousness as being a public relations campaign by the subconscious to make us feel as if we have "something" under control.

Consider the horse who has had a bad experience with a trailer where he banged his head, became frightened, and now no longer wants to put his head in the trailer. He goes precisely up the ramp to that point, then shoots out again backward, even though he didn't bang his head. It can be the same for us. We have made five attempts at loading the horse, and now some other trainer appears who says it's time for us to stop fooling around with this horse and letting him get away with his behavior. We start to feel unsure of ourselves, and we give up, our lack of self-esteem telling us at that moment that we didn't do a good job with our horse in the past, and that's why loading won't work this time. In the worst case, we might even let the trainer who made us nervous take over as we watch as our horse is forced, maybe even beaten into the trailer.

That isn't us—the cognitive person. It's mechanisms that are at work subconsciously. It's the same for the horse when he associates loading with banging his head, being hit with the whip, and loud people shouting at him. Next time, he will start dancing around nervously 30 feet from the trailer.

When somebody says to you, "You mustn't let him get away with that!" practice self-reflection. You are the one who decides what happens with your horse!

Stop the Scenario! Think! Observe! Plan!

Pause, reflect, observe your horse, observe yourself and others, and then quietly plan and consider what you want for you and your horse. Find somebody who can analyze the situation with you. Be self-reflective if, for example, your horse is ignoring your aids and directives. Science-based equine communication shows you how you can teach your horse anything you want.

If you want your horse to halt when you stop while leading him in hand, teach him that with EBEC. Use stimuli and reinforcers to achieve your goal.

Reflect if you have stopped cantering your horse in the arena because he is decoding the leg aids differently, won't go beyond trot, and you don't want to use a whip. Don't just start riding in walk and trot and "live with the situation." Don't avoid your favorite trail because your horse always bolts or spins at the big oak tree. EBEC makes situations like this a thing of the past.

Self-knowledge is the first correct and very important step in this situation! Everything that your horse does is either trained or instinctive behavior. If you notice yourself putting up with things or doing things that don't improve your situation with the horse, the first step is to admit to it—consciously perceive it and bring it to the forefront of your awareness.

Stop what you are doing and start focusing on the solution with EBEC. It has been a key element for changing situations in my own life with horses, with the wonderful side effect that I take it with me into my professional and private life.

SUMMARY
EBEC Level 5

- When all of the levels of the EBEC Pyramid are applied correctly and in the correct sequence, the result is almost magical and works with all horses.
- If we handle horses from their perspective in a way that they can understand and take into consideration all of the scientifically proven facts, horses will be able to fulfil our goals.
- Nothing that we ask of horses corresponds to their natural impulses. However, they can learn everything that corresponds to the characteristics of their breed.
- EBEC helps us to achieve our goals and to be safer with horses.
- Working with EBEC makes us more reflective, more perceptive, and gives us better communication skills, with people as well as with horses.
- Application is a question of practice.

Afterword

In my everyday work with people and horses, I am filled with humility and gratitude, and always a little worried about whether what I am doing will work with the next horse. Sometimes I can't believe it myself but now, thousands of horses later, I have never been disappointed. I have always been able to find the answer in EBEC.

There are always situations in our instructional courses when I am working with participants who have brought difficult horses with them that sometimes make me unsure. As I've described in these pages, I try to determine the appropriate rewarding stimulus from the horse's instinctive behavior. If the horse's behavior doesn't improve after three repetitions, I hear my powerful subconscious say to me: "Perhaps we haven't found a method that works with every horse, after all?"

Then I almost want to turn to any old impulse—maybe send the horse onto a circle or leave him alone in the indoor arena until he calms down. Sometimes the owners suggest getting an extra tool like a whip or auxiliary reins to persuade the horse to give the desired response. I always stop these thoughts.

I gather my closest team of close friends and trainers around me: Melanie Stahr-Herzau and Annika Dethlefs, who teach in the AKA and who are Equine

Coaches, trained and certified according to EBEC. They have an in-depth understanding of the method. We then discuss, analyze, reflect on ourselves and check whether there could be something with this horse that we have overlooked. We think over the learning theories for the hundredth time, sure that the solution is there and that we just need to find the right pathway.

We go through the EBEC Pyramid. Have we forgotten something? Is the horse receptive? We check whether he is able to decode gestures and signals, we go through what the objective is, what he needs to be able to do, and which instinctive reinforcers we are able to identify.

And we stay true to ourselves. The answer is in EBEC. We don't lose sight of that. "We just need to be more creative. We will find the solution—there will be a way for this horse to understand."

It's great to have a team to think things over with, that is ready to reflect, and that doesn't look for the fault in the horse. A highly motivated team that is contactable 24 hours a day. None of what I do would be possible without this creative exchange. We read and examine every study that has been published by scientists across the world, derive new theories, and revise our seminar and course documents. The wonderful thing about the great team at AKA is that we are always working on ourselves. We know what we know, and we also know what we don't know.

When I was writing this book, I was unsure about how certain information is transmitted in the horse's brain. Within just 12 hours I had an answer from a neurological expert on how horses' brains process information, and I thought: "Wow, now we've really reached our goal." That is to say, we've reached the goal of no longer having to blame horses for their behavior, no longer having to raise our voices or use aids to punish them, because they are no longer afraid of us.

And now the (R)Evolution in Horse Training can really begin! My team and I are ready. We can now build on a foundation that allows us to adopt the horse's perspective and to look to ourselves for any incorrectly conveyed information or incorrect training. If we remain true to the EBEC Method, then that is where we will find the solution. And that's a promise!

One of our certified EBEC Equine Master Coaches once said to me: "Andrea, I always find that, just because somebody is older or has been involved with horses for many years, it doesn't mean that they are wiser." I'm not saying EBEC is easy. It takes a lot of practice, a lot of willpower, and a desire for a wonderful, innate understanding of horses. But every trainer or groom who works with horses and loves these wonderful animals and wants to make them their life can benefit from it.

Developing EBEC gave me answers to the questions from my childhood. It makes me grateful for this wonderful, sensitive, elegant, and highly intelligent creature who gives us so much. I no longer take anything for granted. I have been humbled by what horses can do. Some horses began their career as a pleasure horse, then became a show jumper, then a dressage horse, then an eventer. A mare might have quickly had to become a broodmare for a year because her owners wanted a foal, and when that wasn't enough, she might have become a pleasure horse who was then retrained as a therapy horse at the age of 30. That's a lot of learning through life. I don't know whether she would have chosen this path had she been able to choose. But I do know that this path would certainly have caused the horse a lot of anxiety over the years that she would have had to deal with by herself. This is why I have the utmost respect for what horses are capable of as our partners.

My wish for the future is that everybody who has read this book would treat their horses better and more considerately.

Please reach out to me and my team as soon as you have a question. We are ready and we can hardly wait to hear what you think, to have constructive discussions, and to keep developing EBEC further, together with you.

We've made a start. Now it's up to you! I can't wait!

About Andrea Kutsch Academy (AKA)

The (R)Evolution in Horse Training

The Andrea Kutsch Academy (AKA) is the world's leading educational institution for science-based equine communication. It is the only institution that teaches the Evidence-Based Equine Communication (EBEC) training method that Andrea Kutsch has developed over the course of more than a decade in cooperation with international scientists. EBEC is a unique method of working with horses that is based on scientific findings from the horse's perspective; it takes horse training to a whole new level.

AKA offers riders, equestrian professionals, and anyone interested in horses an introduction to EBEC—from one- and two-day seminars to courses over several days. Course content is adapted to participants' prior knowledge so there is something for everyone. By offering certification as "EBEC Trainer" and "EBEC Master Trainer," AKA contributes to spreading new professional standards in the equestrian industry. This means that all equestrian professionals, horse owners, and horses will benefit from the new knowledge.

For more information visit:
andreakutschacademy.com

Sources

Deutsche Reiterliche Vereinigung e. V.: *Richtlinien für Reiten und Fahren.* Vol. 2: Ausbildung für Fortgeschrittene. 12th edition, FN- Verlag der Dt. Reiterlichen Vereinigung, 1997.

Deutsche Reiterliche Vereinigung e. V.: Düe, M. (Ed.): *Die deutsche Reitlehre. Das Pferd,* 1st edition, FN- Verlag der Dt. Reiterlichen Vereinigung, 2002

Deutsche Reiterliche Vereinigung e. V.: Elsner, C.; Kaspareit, T. (Ed.): *FN-Handbuch Lehren und Lernen im Pferdesport.* Warendorf, FN- Verlag der Dt. Reiterlichen Vereinigung, 2007

Hünersdorf, Ludwig: *Anleitung zu der natürlichsten und leichtesten Art Pferde abzurichten, Documenta hippologica.* 2nd Reprint edition of the 1800 Marburg edition. Hildesheim, Olms Verlag, 1992

Heckhausen, J. u. Heckhausen, H. (Ed.): *Motivation und Handeln,* Springer-Lehrbuch. 5th, revised and extended edition, 2018. Springer, Berlin Heidelberg, 2018.

Instruction zum Reit-Unterricht für die Kavallerie vom 31. August 1882. T. 2nd Reprint No. 1-46 incl., Voss, Berlin, 1906.

Klingel, H.: Das Verhalten der Pferde (Equidae). In: Helmcke, J.-G., St- arck, D., Wermuth, H. (Ed.): *Handbuch der Zoologie,* vol. 8. Verlag de Gruyter, Berlin, New York, 1972.

Maslow, Abraham H.: Classics in the History of Psychology, *A. H. Maslow A*

Theory of Human Motivation.

Ödberg, F.O., Bouissou, M.F.: The development of equestrianism from the baroque period to the present day and its consequences for the welfare of horses. *Equine Vet. J.* 31, 1999.

Planet Schule I: Reihe: Das automatische Gehirn [Planet Schule—multimedia school television

from SWR and WDR].

Schäfer, M.: Die Sprache des Pferdes. Nymphenburger Verlagsbuchhandlung, Munich, 1978.

Schäfer, M.: *Beobachtungen zum Verhalten des südiberischen Primitivpferdes* (Sorraiapferd).

Diss. med. vet., Munch, 1986 Sonntag, I. (Ed.): H.Dv. 12, *Die deutsche Reiterklassik hat einen Namen*, 1937.

Schondorf, Wu-Wei-Verlag 2008 Van de Poll, N. E.; van Goozen, S. H.: Hypothalamic involvement in sexuality and hostility: comparative psychological aspects. In: *Progress in Brain Research*, vol. 93, p. 343–361, 1992.

Acknowledgments

I am grateful to George H. Waring, Emeritus Professor of Southern Illinois University in Carbondale, Illinois, for his support and fundamental research into the body language of horses. The ethograms, such as those included in this book, enable us to meaningfully connect the nonverbal and intra- and interspecies communication of horses with the learning theories of the EBEC Pyramid.

Index

Page numbers in *italics* reference figures.

Acoustic communication in horses, 72–79, 96
Adrenal gland and adrenaline, 48, 50–51, 60
Affiliation motivation, 36
Age and aging, 61–62
Agonistic (aggressive) behaviors
    expressions of aggression, 92, 94, *100,* 108, *109, 116,* 117
    hitting and shouting as, 71–72
    intermittent reinforcement for, 147–149
    learning and, 106
    survival instincts and, 38–39, 158
    threshold for, 157–158
AKA (Andrea Kutsch Academy), 11–12, 79, 161, 166–167
Alarm
    expressions of, 73–74, 96, 106–107, *107*
    threshold for, 157
Amygdala, 48–49, 51, 56, 58
Andrea Kutsch Academy (AKA), 11–12, 79, 161, 166–167
Army Riding Regulation 12: German Cavalry Manual on the Training of Horse and Rider, 10
Attention of horse
    directed forward, 101
    directed to rear, *100,* 103–106, *103*
    directed to side, 101–102, *102*
Autonomic nervous system, 65
Aversive stimuli
    defined, 132
    from the horse's perspective, 136–139
    negative reinforcement and, 134–135

Bachelor of Science in "Horse Communication, Riding, Training, and Teaching," 11
Backward-directed perception, *100,* 103–106, *103*
"Bad" behavior, 29–30, 44–45
Basic needs, 34, 35, *36,* 37, 38. *See also* Physiological and psychological needs
Behavior, communication and, 125–128
Behavioral biology (ethology), 22
Behaviorism, 126
Blood pressure, 48–49, 65
Blow (acoustic signal), 73–76
Body language, of horses. *See* Ethograms
Body temperature, 53
Breaks, as rewarding stimuli, 123

Chewing and licking, 65–67, 112–115, 119, 133
Chronic stress, 51–54, 61
Classical conditioning, 20–21, 32–33, 128, 129–131, 157
Classical equestrianism, 13–19, *14*
Clicker training, *14,* 18, 21
Coercive aids
    author's first introduction to, 2
    current use of, 23, 24, 26
    historic use of, 15–18, 31, 32
    "as means to an end," 18
    as punitive stimuli, 17–18, 123
    stress and learning, 56–57
    suppression of expressive behavior with, 108–111
    violence-free communication versus, 24
    voice as, 17
Comfort zone, 64–65, 66, 133, 138–139, 141
Communication
    author's early experience, 1–8
    behavior and, 125–128
    body language, of horses, 40–42. *See also* Ethograms
    decoding and encoding gestures and signals, 8–9, 69–72, 81–88
    defined, 68, 125–126
    evolution of, 10, 13–33. *See also* Evolution of horse training
    FN (German National Equestrian Federation) on, 40–41
    horse whispering and, 8–9
    learning theories on, 125–128
    scientific basis for, 11–12. *See also* Evidence-Based Equine Communication (EBEC) pyramid
    types of, 68–69. *See also* Acoustic communication in horses; Interspecies communication; Intraspecies communication; Nonverbal communication; Violence-free communication
Conditioned, positive reinforcer, 147–149

Conditioning, 128–136
    classical conditioning, 20–21, 128, 129–131, 157
    defined, 82–83
    operant (instrumental) conditioning, 17–18, 20, 126, 128, 131–136, 157
    predator perspective conditioning, 20–22
Cortisol, 51–52, 58

Decoding and encoding gestures and signals, 8–9, 69–72, 81–88
Dethlefs, Annika, 166–167
Die natürlichste und leichteste Art Pferde abzurichten [The Easiest and Most Natural Way to Train Horses] (Hünersdorf), 14–15, 122
Dogs, 20–21, 32–33
Doing the right thing, 50–51
Dominik (pony), 5–6, 28, 160

Ear positions, 72, 99, *100,* 101, 106–107, *107,* 108
Empathy, 161
Encoding and decoding gestures and signals, 8–9, 69–72, 81–88
Equestrianism (classical), 13–19, *14*
Ethograms, 40–42, 84–118
    about: overview and summary, 22, 40–42, 43, 49, 84–89
    aggression expressions, 92, 94, *100,* 108, *109, 116,* 117
    defined, 40
    facial expressions, 98–108, *100, 102–103, 107, 109*
    full body expression gestures, 91–92
    impressive gestures, 96–98
    learning theories on, 43, 49, 125–128
    leg and hoof gestures, 92–96, *93*
    positive reinforcements and, 133
    reading ethograms, 89–91
    sensory well-being expressions, 111–112, *113*
    snapping (licking and chewing), 65–67, 112–115, 119, 133
    suppression of expressive behavior, 108–111
    tail gestures, 115–118, *116*
Ethology (behavioral biology), 22
European University Viadrina, 11
Evidence-Based Equine Communication (EBEC) pyramid
    about: overview and summary, 10, 11–12, *14,* 29–33, *36*
    level 1, 34–67. *See also* Physiological and psychological needs
    level 2, 68–121. *See also* Interspecies communication
    level 3, 122–156. *See also* Learning theories
    level 4, 157–159
    level 5, 160–165
Evolution of horse training, 13–33
    about: overview and summary, 10–12, *14*
    stage 1: classical equestrianism, 13–19
    stage 2: predator perspective conditioning, 20–22
    stage 3: natural horsemanship and horse whispering, 22–28. *See also* Natural horsemanship and horse whispering
    stage 4: EBEC, 29–33. *See also* Evidence-Based Equine Communication (EBEC) pyramid
"Experienced horseman," 41
Expressions and gestures. *See* Ethograms

Facial expressions, 98–108, *100, 102–103, 107, 109*
Field of vision, 56–59, *57,* 72, 104
FN (German National Equestrian Federation), 40–41, 118–121
Foal play, *95*
Focus and objectives, of EBEC, 157–159
Foreleg gestures, 92, 94–96
Forward-directed attention, 101–102, *102*
Freie Universität, 11
Fulfilling potential, 160–165
Full body expression gestures, 91–92

Gadgets, 55. *See also* Coercive aids
"German Horse Whisperer," 9, 29
German National Equestrian Federation (FN), 40–41, 118–121
Gestures and expressions. *See* Ethograms
The Gymnasium of the Horse (Steinbrecht), 13

H. Dv. 12: Army Riding Regulation 12: German Cavalry Manual on the Training of Horse and Rider, 10, 13, 18–19
Health (physical and mental), 54–56. *See also* Physiological and psychological needs
Heart-rate monitors, 48
Heckhausen, Heinz, 39
Hierarchy of needs, 34, 35–39. *See also* Physiological and psychological needs
Hind leg gestures, 92, *93,* 94
Hippocampus, 56–58
Hitting. *See* Coercive aids
Hoof gestures, 92–96, *93*
Horse body language. *See* Ethograms
"Horse Communication, Riding, Training, and Teaching" degree, 11
Horse racing, 29–31
Horse's point of view. *See* Evidence-Based Equine Communication (EBEC) pyramid

Horse training evolution. *See* Evolution of horse training
The Horse Whisperer (film), 9, 22
Horse whispering. *See* Natural horsemanship and horse whispering
Humility expressions, 112, 114, 115, *116*
Hünersdorf, Ludwig, 14–16, 42, 122
Hypothalamus, 47–48, 51

Impressive gestures, 96–98
Incidental learning, 127
Innate versus learned behavior, 139–156
    about: overview and summary, 139–141, 143
    instinctive reinforcement and, 141–145, 149
    positive reinforcement and, 141, 144, 147–149
    punitive stimuli and, 145–147, 157–159
    sensory memory, 149–154
    timing and scope for interpretation, 154–156
Instinctive reinforcement, 141–145, 149
Instinct theory and behavior, 20, 21, 22, 140
Instruction zum Reit-Unterricht [Instructions for riding lessons] (Stritter), 18
Instruction zum Reit-Unterricht für die königlich preußische Kavallerie [Riding Instruction for the Royal Prussian Cavalry], 13–14
Instrumental (operant) conditioning, 17–18, 20, 126, 128, 131–136, 157
Intentional learning, 127
Intermittent reinforcement, 147–149
Interpretation, timing and scope for, 154–156
Interspecies communication, 68–121
    about: overview and summary, 68–69, 121
    acoustic communication in horses, 72–79
    behavior and, 125–128
    conscious processes, 80–81
    decoding and encoding gestures and signals, 8–9, 69–72, 81–88
    defined, 68, 69
    EBEC training plans and, 79
    horse's expressive behavior, 69–72, 84–118. *See also* Ethograms
    horse's response to, 83–84
    learning theories on, 43, 125–128
    natural horsemanship and, 25–26
    nonverbal communication, 22–28, 80–83
    Training Scale application, 118–121
"Into-pressure syndrome," 145
Intraspecies communication, 68–69, 85, 97–98

Key stimuli, 139, 143
Kicking, 92, *93*, 94
Knocking, 92, *93*, 94

Kutsch, Andrea
    Andrea Kutsch Academy (AKA), 11–12, 79, 161, 166–167
    background, 1–2
    competition path, 5–8
    EBEC foundation, 10, 11–12
    as "German Horse Whisperer," 9, 29
    on gratitude and humility, 166–168
    horse whispering "inspiration," 8–9
    inner conflict, 4–5
    longing for unity with horses, 2–4
    polo and, 7–8
    research background, 11–12

Labeling horses as "bad," 29–30, 44–45
Learned versus innate behavior. *See* Innate versus learned behavior
Learning theories, 122–156
    about: overview and summary, 122, 156
    acoustic signals and, 78–79
    on classical conditioning, 20–21, 32–33, 128, 129–131, 157
    comfort zone and, 64–65, 66
    on communication and behavior, 43, 125–128
    innate versus learned behavior, 139–156. *See also* Innate versus learned behavior
    on operant conditioning, 17–18, 20, 126, 128, 131–136, 157
    physiological needs and, 37–38
    problematic training stimuli, 122–124
    on reward and punishment, 37–38, 124–125, 136–139
    stress and learning, 32–33, 37–38, 43, 63–65
    timing and, 106
    transforming learning into behavior, 159
Leg gestures, 92–96, *93*
Lewitz Stud, 12, 61–62, 66–67, 154
Licking and chewing (snapping), 65–67, 112–115, 119, 133
Lifting the hind leg, 92
Limbic system, 57–58
Lip gestures, 98, *100*, 112. *See also* Mouth gestures; Snapping
Long-term memory, 25, 57, 59, 127, 150, 152–153
Lorenz, Konrad, 20

Maslow, Abraham, 34, 35–39
Memory, of horses
    about, 59–61, 64
    long-term memory, 25, 57, 59, 127, 150, 152–153
    short-term (primary) memory, 57, 127, 150–151, 153

Mental health, 54–56
Mirror neurons, 152
Motivation, 39, 159
Motivation and Action (Heckhausen), 39
Mouth gestures, 99, *100*, 106–107, *107*, 112
Multitasking, 157–158
Muscle tone and tension, 48–49, 50–52, 54, 55

Natural horsemanship and horse whispering
    about, *14*, 18, 22–28
    foundation of, 14
    "German Horse Whisperer," 9, 29
    "inspiration" from, 8–9
    interspecies communication and, 25–26
    licking and chewing as sign of force with, 119
    punishment and, 16–17
Needs. *See* Physiological and psychological needs
Negative reinforcement, 17–18, 27, 131–132, 134–136
"Neutral state" behavior, 90
Neutral stimulus, 20–21, 129–131
Nonverbal communication, 22–28, 80–83
Nostril gestures, 99, *100*, 101, 106–107, *107*, 108, 112

Objectives and focus, of EBEC pyramid, 157–159
Object play behavior, *95*
Operant (instrumental) conditioning, 17–18, 20, 126, 128, 131–136, 157

Pain, 24, 55, *100*, *116*, 123. *See also* Coercive aids
Parasympathetic nervous system, 65
Pavlov, Ivan Petrovitch, 20–21, 32–33, 128
Pavlovian conditioning, 32–33, 128. *See also* Classical conditioning
Pawing, 92, *93*, 94, 96
Perception, of horses, 56–59, *57*, 72, 104
Physical health, 54–56
Physical stressors, 75
Physiological and psychological needs, 34–67
    about, 34–35
    about: overview and summary, 67
    age and aging, 61–62
    chronic stress, 51–54, 61
    doing the right thing, 50–51
    ethograms and, 40–42
    health (physical and mental), 54–56
    horses are not inherently "bad," 44–45
    horses don't scheme and strategize, 35, 39, 45–46, 135
    horse's point of view, 42–51
    hypothalamus and amygdala, 47–49, 51, 56, 58
    memory of horses, 59–61, 64
    perception of horses, 56–59
    pyramid of, 35–39
    responding with understanding, 46–47
    stress hormones and, 32–33, 39, 43, 44–45, 48
    stress management, 38, 51–52, 62–67
    stress measurement, 52–54
Piaffe, 97, 105
Play behavior, 94, *95*
Pleasant stimuli, 136–139
Pleasure, state of, 137–139
Point of view of horses. *See* Evidence-Based Equine Communication (EBEC) pyramid
"Poling," 105
Polo, 7–8
Positive reinforcement, 131–132, 133–134, 141, 144, 147–149
Potential, fulfilling of, 160–165
Predator gestures, 87, 129
Predator perspective, 10, 20–22, 32–33
Prefrontal cortex, 55
Pressure
    classical conditioning and, 129–131
    coercive aids as, 15–18
    as conditioned, positive reinforcer, 148
    ethograms for assessment of, 91, 104
    pressure scale timing and scope, 155–156
    "into-pressure syndrome," 145
    as punitive stimuli, 123, 146
    sensory memory and, 152–154
Prey perspective, 32–33
Primary (short-term) memory, 57, 127, 150–151, 153
Primary reinforcers, 124–125
The Principles of Riding (FN), 24
Prioritization, of needs, 39
Proprioceptive communication, 68
Psychological stressors, 75
Pulse (heart rate), 48, 51, 53–54, 55, 65, 107
Punishment, 16–19, 71, 145–147
Punitive stimuli, 123, 145–147, 157–159
Pyramid. *See* Evidence-Based Equine Communication (EBEC) pyramid

Race horses, 29–31
"Rapping," 105
Rear-directed attention, 103–106, *103*
Redford, Robert, 9
Reflective behavior, 140
Refusals, 42–43, 54
Reinforcement
    in classical conditioning, 130–131
    conditioned, positive reinforcer, 147–149
    instinctive reinforcement, 141–145, 149
    intermittent reinforcement, 147–149

learning theories, 140–141
negative reinforcement, 17–18, 27, 131–132, 134–136
in operant conditioning, 131–136
positive reinforcement, 131–132, 133–134, 141, 144, 147–149
primary reinforcers, 124–125
secondary reinforcers, 124–125
Respiratory rate, 51, 52, 53–54, 55, 107
Restraining horses, 63
(R)evolution of horse training. *See* Evolution of horse training
Reward and punishment
instinctive reinforcement and, 142
learning and, 37–38, 122–125, 136–139
punitive stimuli, 123, 145–147, 157–159
rewarding stimuli, 123–124, 157–159
reward system, scientific basis for, 15

Saddling gestures, 94–96
Safety and security needs, 35
Saliva production, 65–66
Scheming and strategizing, by horses, 35, 39, 45–46, 135
Schockemöhle, Paul, 11–12, 61–62, 154
Scope, for interpretation, 154–156
Scratching, *93*, 94, 111–112, *113*
Secondary reinforcers, 124–125
Security and safety needs, 35
Seductress (horse), 29–32, 34
Self-realization, 36
Self-reflection, 50, 78, 140, 162–164, 166–167
Sensory memory, 149–154
Sensory perception, 141
Sensory well-being expressions, 111–112, *113*
Short-term (primary) memory, 57, 127, 150–151, 153
Sideways-directed attention, 101–102, *102*
Skinner, Burrhus Frederic, 20
Snapping (licking and chewing), 65–67, 112–115, 119, 133
Snore (acoustic signal), 77–79
Social needs, 36–37
Speech, 17, 72–73
Spooking, 44–45, 96, 139, 143–144
Spurs. *See* Coercive aids
Stahr-Herzau, Melanie, 166–167
Stamping, 92, *93*, 94
"Stamp that the horse has on his forehead," 29–30
State Stallion Depot of the Stud, 26–27
Steinbrecht, Gustav, 13

Strategizing and scheming, by horses, 35, 39, 45–46, 135
Stress
amygdala's response to, 48–49
chronic, 51–54, 61
learning and, 32–33, 37–38, 43, 63–65
management of, 38, 51–52, 62–67
measurement of, 52–54
responses to, 74–75
snapping (licking and chewing) and, 65–66, 112, 114–115
types of, 75
Stress hormones, 32–33, 39, 43, 44–45, 48
Striking, 92, *93*
Submission, 15–17, *100*
Sundance (horse), 7–8
Suppression of expressive behavior, 108–111
Survival instincts, 38–39, 43, 158
Symmetrical communication, 69
Sympathetic nervous system, 65

Tail carriage and gestures, 115–118, *116*
Taking breaks, as rewarding stimuli, 123
Team collaboration, 166–167
Teeth grinding, 55
Thorndike, Edward Lee, 20, 137–138
Thoroughness, 118–119
Timing, for interpretation, 154–156
Tinbergen, Nikolaas, 20, 142
Touch, as rewarding stimuli, 123
Training Scale (FN), 118–121, 158–159
Training skills, 157–159
Treats, as rewarding stimuli, 123
Turniertrottel "TT" (competition groom), 7

Unconditioned response, 130–131
Unconditioned stimulus, 130–131
Unidirectional communication, 69
University of Zurich, 11

Verbal language, 17, 72–73
Violence-free communication, 24
Vision of horses, 56–59, *57*, 72, 104
Voice, use of, 17, 72–73, 123

Well-being expressions, 111–112, *113*
Whips. *See* Coercive aids
Withdrawal, as punitive stimuli, 123, 146